FIRST IN THE WORLD

THE STOCKTON & DARLINGTON RAILWAY

John Wall

SUTTON PUBLISHING

First published in 2001 by
Sutton Publishing Limited · Phoenix Mill
Thrupp · Stroud · Gloucestershire · GL5 2BU

Copyright © John Wall, 2001

All rights reserved. No part of this publication may be reproduced, stored in a retrieval system, or transmitted, in any form, or by any means, electronic, mechanical, photocopying, recording or otherwise, without the prior permission of the publisher and copyright holder.

John Wall has asserted the moral right to be identified as the author of this work.

British Library Cataloguing in Publication Data
A catalogue record for this book is available from the British Library.

ISBN 0-7509-2729-1

Typeset in 10/12 pt Times.
Typesetting and origination by
Sutton Publishing Limited.
Printed and bound in England by
J.H. Haynes & Co. Ltd, Sparkford.

Contents

	List of Illustrations	v
	Acknowledgements	vii
	Abbreviations	viii
	Introduction	ix
Chapter 1	PRELIMINARIES: FROM PROVINCIAL POLITICS TO ACTS OF PARLIAMENT	1
Chapter 2	FOUNDATIONS: ENGINEERING	18
Chapter 3	THE GRAND OPENING	43
Chapter 4	THE EARLY YEARS: PRIVATE RISK FOR PUBLIC SERVICE	58
Chapter 5	REAL ESTATE AND CUSTOMER SERVICES	82
Chapter 6	THE RISE OF THE RAILWAY INDUSTRY	98
Chapter 7	MAIN-LINE BRANCHES AND EXTENSIONS	108
Chapter 8	NEW LINES: THE BUILDING OF AN EMPIRE	124
Chapter 9	RIVAL RAILWAYS	143
Chapter 10	DARLINGTON – FIRSTBORN OF RAILWAY TOWNS: A PORTRAIT	154
Chapter 11	HERITAGE	165
	Appendices	179
	Notes	191
	Bibliography	205
	Index	208

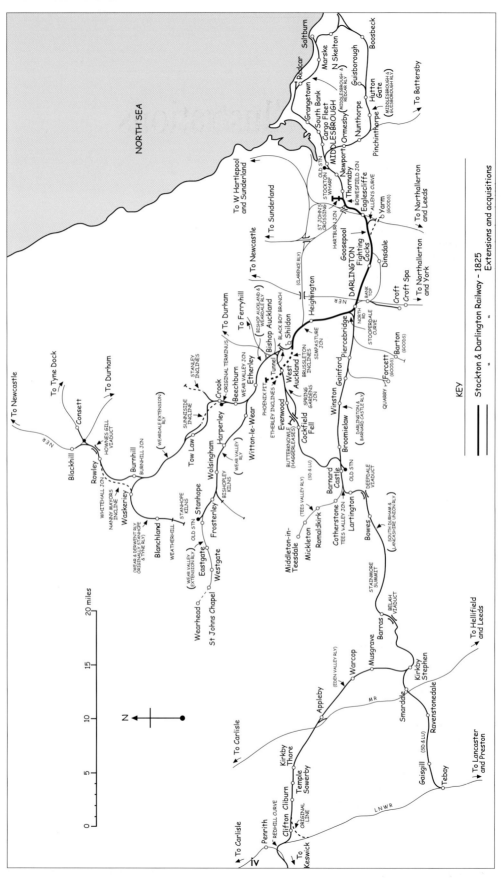

The Stockton & Darlington railway network in 1863

List of Illustrations

Black & White Illustrations

Page

iv	The Stockton & Darlington railway network in 1863
2	Plan of proposed canal from Stockton via Darlington to Winston, 1772
10	George Stephenson, by Henry Briggs
15	The S&DR seal and motto
22	The Gaunless Bridge
34	An S&DR share certificate
39	George Stephenson's Killingworth engine, 1816
40	*Locomotion No. 1* at Darlington Bank Top station
52	The Opening of the S&DR. Sketch by John Dobbin
54	The Opening of the S&DR. Drawing by J.R. Brown
55	The inaugural train crossing Skerne Bridge (unknown artist)
60	The first railway timetable
62	The first railway coach, *Experiment*
63	*Experiment*, as illustrated in Theodore West, 1885
68	Early private railway coach, *The Union*
70	Daniel Adamson's railway coach-house, Shildon
72	Chauldron wagon preserved at the National Railway Museum
73	Dandy cart at the National Railway Museum
77	Timothy Hackworth's locomotive, *The Royal George*
78	*The Globe* at the head of Timothy Hackworth's business card
81	An S&DR milepost *in situ*
84	The birthplace of Edward Pease – The Grove, Darlington
84	Edward Pease's home, 146 Northgate, Darlington
85	George Stephenson instructing the daughters of Edward Pease
88	The Fighting Cocks Inn, Darlington
89	Commemorative plaque on the preserved S&DR booking office
90	An early S&DR paper ticket, issued in 1835
96	The first railway ticket office and goods warehouse
97	North Road station, Darlington, as it appears today
98	The Mason's Arms, Shildon, site of the inugural run
100	Timothy Hackworth's home, Soho Works, Shildon
104	The pattern shop and storage shed, Soho Works, today
106	Locomotive No. 25 Derwent at Darlington Railway Museum
116	World's first railway suspension bridge over the Tees
117	Medal issued to commemorate the Middlesbrough branch
118	Plan of Middlesbrough with the new dock and railway
122	The S&DR's Zetland Hotel, Saltburn, today
141	Sir Joseph Whitwell Pease

146	S&DR crossing, Albert Hill, Darlington, 1925
158	Bank Top Station, Darlington
162	Coat of arms of the Borough of Darlington
170	S&DR early passenger coach, Darlington Railway Museum
172	Statue of Joseph Pease by Joseph Lawson, High Row, Darlington
174	'First in the World' – Centenary medal, 1925

COLOUR PLATES
between pages 116 and 117

1	*The Opening of the Stockton & Darlington Railway* by John Dobbin
2	Edward Pease, 'Father of Railways', by Heywood Hardy
3	Joseph Pease, by James MacBeth
4	The family of George Stephenson, by John Lucas
5	Robert Stephenson, by John Lucas
6	Richard Trevithick, by John Linnell
7	George Hudson, 'the Railway King', by Sir Francis Grant
8	*Locomotion No. 1* races a stagecoach, by Terence Cuneo
9	Portrait from the life of Timothy Hackworth (unknown artist)
10	Trevithick's *Pen-y-Daren* locomotive, 1804, by Terence Cuneo
11	High Street House, Wylam, birthplace of George Stephenson, by Ronald Embleton
12	The inaugural train crossing Skene Bridge (unknown artist)
13	*Locomotion No. 1* near the Stockton terminus, by John Wigston
14	The King's Head Inn, Darlington, by William Dresser
15	Etherley incline, 1875 (unknown artist)
16	Brusselton incline and engine house (unknown artist)
17	'First in the World': poster designed by Andrew Johnson

Acknowledgements

One of the most rewarding aspects of the preparation of a work such as this is the opportunity (or excuse!) to travel, visiting a host of museums, libraries and archives along the way. However, it is only in retrospect that one realises just how much one is indebted to others for their patience and forbearance, and how many officers of those institutions have gone out of their way to bring to light often elusive material. Without exception, staff have been courteous and obliging, and if most have to remain anonymous my gratitude is no less sincere for that. In particular, though, I would like to express my warm appreciation to the following, without whose ready help and kindly interest this history could not have been completed.

My first thanks go to the staff of the National Railway Museum's Library and Archive in York, and of the National Museum of Science and Industry's Library in London, and especially to Beverley Cole and Wendy Sheridan, Curators of the Pictorial Collections at York and London respectively. In this context, the Information Department and the Heintz Archive and Library of the National Portrait Gallery in London have also been particularly helpful.

Nearer home I commend the up-to-date facilities and the professionalism of the staff at the Durham County Record Office, County Hall, Durham City, and the North Yorkshire County Archive and Record Office at Northallerton. Two libraries in particular have proved a mine of information through the competence of their dedicated staff: the Darlington Centre for Local Studies, a Department of Darlington Library; and Durham City Central Reference Library, where the digitalised 'Durham Record of Historic Maps and Photographs' has been invaluable.

We are particularly fortunate to possess in this north-east region so many museums of high quality, which together represent a rich repository of railway history. Foremost among them, of course, is the National Railway Museum which has assumed greater relevance with the transfer there of a number of exhibits from the National Museum of Science and Industry, resulting from a reorganisation of its transport displays. Equally important are two local museums, both with national associations. First, the Darlington Railway Centre and Museum at North Road Station, imaginatively cared for by the Curator, Steven Dyke. It is highly appropriate that the Study Centre for North East Railway History named for Ken Hoole (1916–88) should be housed in the same building, and I owe a particular debt to Ann Wilson, Darlington Museum's Documentation Officer and who looks after the Centre, for making available material that could be found nowhere else. Second, the Timothy Hackworth Victorian and Railway Museum at Shildon has attracted many rewarding visits over the years, and I take this opportunity to thank the Curator, Alan Pearce, for sourcing original material on the Black Boy branch and Stockton and Darlington Railway Company Coaches.

I know that those already named would not think it amiss if, finally, I pay a special tribute to two dedicated individuals who are now retired: Peter W.B. Semmens, formerly Deputy Director of the National Railway Museum, has kindly reviewed and wisely advised on the section that deals with inclined planes, as has Alan Suddes, formerly Curator of the Borough of Darlington Museum and Art Gallery Collections, on the section that deals with John Dobbin's painting *The Opening of the Stockton and Darlington Railway*.

All these good people have made an indispensable contribution to what has been, for me, a labour of love: a belated tribute to my native town, and its unique place in railway history.

It could be objected that a 'new' account of the S&DR is a gratuitous undertaking in the light of the many books that have already been published. It is true that a great deal has been written about the S&DR, and that in itself is a recognition of the importance of the subject. The one drawback of this earlier material is that it is not for the most part readily accessible to the general reader, either because it forms part of more comprehensive works, or because it deals with a specialised aspect of the subject. A notable example of the former problem is W.W. Tomlinson's monumental 816-page volume *The North-Eastern Railway: its rise and development*, first published in 1915 but substantially completed by 1905, which has been a mine of information for writers in this field ever since. Although it 'includes a detailed history of the S&DR', this is scattered throughout the book rather than available in one compact section. Similarly, L.T.C. Rolt's definitive biography *George and Robert Stephenson: The Railway Revolution*, first published in 1960, devotes one excellent chapter to the S&DR, but beyond that, the volume has to be trawled for other relevant material. A scholarly and indispensable example of a more specialised work is *The Origins of Railway Enterprise: the Stockton and Darlington Railway, 1821–1863*, by Maurice W. Kirby, Reader in Economic History at Lancaster University, which was published as recently as 1993. Another up-to-date, but rather technical work is T.R. Pearce's *The Locomotives of the Stockton and Darlington Railway*, published in 1996. It will be evident from the text, footnotes and bibliography how heavily the present author is indebted to these, and to other writers, to whom due acknowledgement is accorded by one who is obliged to 'stand on the shoulders of giants'.

Abbreviations

CK&PR	Cockermouth, Keswick & Penrith Railway
D&BCR	Darlington & Barnard Castle Railway
EVR	Eden Valley Railway
GNER	Great North of England Railway
L&CR	Lancaster & Carlisle Railway
L&NER	London & North Eastern Railway
L&NWR	London & North Western Railway
LMR	Liverpool & Manchester Railway
LNR	Leeds Northern Railway
M&GR	Middlesbrough & Guisborough Railway
M&RR	Middlesbrough & Redcar Railway
N&DJR	Newcastle & Darlington Junction Railway
NER	North Eastern Railway
S&DR	Stockton & Darlington Railway
S&TR	Stanhope & Tyne Railway
SD&LUR	South Durham & Lancashire Union Railway
WDR	West Durham Railway
Y&NMR	York & North Midland Railway
YN&BR	York, Newcastle & Berwick Railway

Introduction

When in 1925 the London & North Eastern Railway (the L&NER) began to plan the celebrations which were to mark the centenary of the opening of the Stockton and Darlington Railway (the S&DR), the organisers appreciated the need for a slogan that would capture the imagination of the public. What they came up with was simple and appropriate – FIRST IN THE WORLD. Initially it was used on a poster that announced 'George Stephenson Started Us 100 Years Ago. L.N.E.R. First in the World'. Of course, this was a simplification of a complex story, as George Stephenson, whose portrait graces the rest of the poster, would have been the first to admit. Nevertheless, the S&DR, the predecessor of the L&NER, was indeed the 'First in the World', and the slogan was quite rightly repeated during the 1925 centenary celebrations. The excited schoolchildren who travelled by the 'Flying Scotsman' train in the procession that was the highlight of the celebrations, for example, were issued with a special ticket which bore the legend 'L.N.E.R. First in the World. 1825–1925: Darlington to Stockton: Stockton to Darlington'. The slogan also featured on a now-rare medal that was issued to commemorate the centenary.

Since the opening of the S&DR, there have been two other historic celebrations – the jubilee in 1875 and the sesquicentenary in 1975. A 175th anniversary celebration, which occurred in the year 2000, might seem to be unnecessary or, comparatively, of much less significance. However, there are compelling reasons why this anniversary should be fittingly observed, not least because the Millennium provides us with a unique opportunity to assess the historic importance of Great Britain's industrial heritage. A major part of that inheritance is the railway industry, which was born in Britain and then quickly spread throughout the world. Because the S&DR was without question the world's first steam-worked public railway, we may fairly claim for it that uncompromising title FIRST IN THE WORLD. Another anniversary celebration to signify the opening of the S&DR in the Millennium year is therefore surely in order. In addition, there is a strong case for emphasising the special circumstances of particular places that have played a leading role in the life-story of the railways. The enduring place of Darlington in the title of the world's first steam-worked public railway exemplifies its priority in that story.

The large number of histories of the S&DR that have been published in times past is an indicator of its importance, and there may seem to be no justification to add to their number. However, there is one aspect of the history of the S&DR which has not yet been explored in detail. That is the crucial relationship of the town of Darlington to its distinguished protégé, and the quite unique nature of Darlington as the world's first railway town of any consequence. The qualification 'of any consequence' is deliberate. Although Shildon does not appear in the company's title, there are those who would insist that *it* was historically the world's first railway town. Yet during the time that Shildon hosted a part of the S&DR it never aspired to be a *town*, in the sense that Darlington always was, or to be a railway *centre*, as Darlington undoubtedly became. Its population was never more than one-tenth that of Darlington. Nevertheless, the world's first railway engineering works were established there, as we shall see, before they were transferred to Darlington. It is because Shildon's fortunes, in railway terms, were so intimately bound up with Darlington that its history is here included in Darlington's history as the 'First in the World'.

At one of the early discussions between the railway engineer George Stephenson and the Quaker financier Edward Pease – each in his own sphere regarded as the 'Father of Railways' – Stephenson pointed out that the shortest line between the collieries and Stockton would be by Aycliffe and not by Darlington. Edward Pease pulled him up, and said with marked emphasis and determination, 'George, thou must think of Darlington: thou must remember Darlington sent for thee.'

By the time of the 50th anniversary celebrations of the S&DR, Darlington's place in the history of railways was already recognised and assured. In his *Jubilee Memorial of the Railway System*, published in 1875, J.S. Jeans can confidently claim:

> As the birthplace of the railway system, Darlington stands unrivalled and alone. It was in Darlington that the scheme was first mooted. It was in Darlington that it was brought to maturity. It was in Darlington that all the opposition to the project was conciliated or conquered. Darlington supplied nearly all the ways and means. Darlington furnished the sinews of war for Parliamentary victory; every commercial and mechanical problem was solved at Darlington; and, in a word, Darlington, and Darlington alone, was the cradle of the system which has now attained its jubilee year, after passing through untold vicissitudes, and ultimately reaching the full stature of robust and vigorous maturity. Three-fourths of the capital, and more than three-fourths of the risk and responsibility were undertaken in and through Darlington. Need more be said, then, to vindicate the claim of that town to be considered the birthplace of the railway.

After making due allowance for the author's Victorian style, that claim is as true today as when it was written 125 years ago. In an address to Edward Pease delivered by Francis Mewburn, Solicitor to the S&DR, in the great man's home in Darlington, to honour his achievements, Mewburn declared: 'To yourself and your active colleagues . . . we owe entirely the advantage of our town being the focus whence sprang the means of locomotion you originated.'

However, Darlington as its centre cannot be divorced from the S&DR system as a whole. The fortunes of the one were intimately bound up with the fortunes of the other. One example of this symbiotic relationship nicely illustrates the point: individual members of the Quaker Pease family of Darlington, without whom the S&DR would not have sprung into being, were partners in commercial enterprises as diverse as coal mines, coke ovens, ironstone mines and limestone quarries, and many others, throughout the area which was served by the S&DR in its maturity. This was a mutually dependent relationship – the railway as a transport artery was crucial to the success of these far-flung commercial interests, and in its turn the railway depended on the revenues that it derived from them.

So FIRST IN THE WORLD has been chosen as the title of this history because it applies with equal force to the S&DR as the world's premier railway, as to Darlington as the world's premier railway town. The description of the S&DR as 'the world's first steam-powered public railway' is carefully chosen. The Stockton and Darlington was not the first railway. For example, there were already wooden railways in the north of England, beginning as early as 1609, mainly for the transport of coal in horse-drawn wagons or 'chauldrons' from the pitheads to staithes on the river banks and estuaries. The combination of terms in this designation is therefore important, for while other railways can claim priority in one or other of these respects, the S&DR was the first to incorporate all those elements that went to make for a truly revolutionary mode of transport worldwide. In particular, it was a *public* railway in the sense that goods belonging to anyone could be transported on payment of a toll, and not just the proprietor's merchandise. It was also a public railway in the sense that any member of the public could be a fare-paying passenger.

It has become a truism that the Industrial Revolution began in Britain before it spread to affect, to a greater or lesser degree, the destinies of all the nations. It is no less true that it was the fortuitous coming-together of a number of discrete factors in Great Britain that sparked the Industrial Revolution. It was also the coming-together of many of those same factors that brought about the birth of that major player the Railway on the stage of the Industrial Revolution, as a cheap and efficient means of transport for goods and passengers. Among those factors were, foremost, the local availability of adequate supplies of coal and iron ore, the development of the iron rail, the application of steam power to locomotive engines, and perhaps most important of all, the conjunction of men with mechanical, inventive genius, and men of commercial vision and entrepreneurial flair.

We cannot underestimate the effect of the Industrial Revolution on the way of life of the greater part of mankind, and in its final decades, in particular, on the lives of the people of the United Kingdom, the cradle of the Industrial Revolution. Whereas before, the horizons of the majority of men and women were limited to the boundaries of their place of birth – village, or town, or valley – because of the obstacles in the way of travelling further afield, henceforth their horizons were limited only by their inability to pay for a common means of transport that was swift, efficient and comparatively cheap. Whereas before, journeys were measured by the distance the poor could walk in a day, or the better-off could ride in a stagecoach or on horseback, henceforth journeys were limited only by the hitherto unimaginable speed a train could travel to all points of the compass. Whereas before, the transport of goods was limited by the load a packhorse could carry between sunrise and sunset, henceforth the carrying of merchandise was limited only by the load that the mighty, many-horsepowered locomotive could haul. It is therefore not trite but true to say that the world took a quantum leap forward with the Industrial Revolution, and that we are still, in many respects, its beneficiaries.

To summarise. A principal element of the Industrial Revolution was the building of the railways. The pioneer and progenitor of the railway system was the S&DR: and the executor and legatee of the S&DR was the town that it was pleased to incorporate in its title. In *this* context it is the author's hope that this history will provide a long-overdue, if not exhaustive, record of the part his native town has played in furthering the development of transport systems worldwide. A history of this kind must necessarily include much by way of description of locations and buildings associated with its subject. Towards the close it will become apparent how much Darlington has lost of its railway inheritance. Yet there is a surprising amount still remaining to be discovered and appreciated by visitors in general, and by the discerning pilgrim to this railway shrine in particular. These physical reminders, which are eloquent of Darlington's biography as the premier railway town, are given special attention in the following pages.

CHAPTER I

Preliminaries: From Provincial Politics to Acts of Parliament

PREPARING THE GROUND: A CONFLICT OF INTERESTS

There was a prolonged period of gestation for the S&DR before its official opening on 27 September 1825. In the preceding years many events, meetings and individuals were to play their part. One man in particular had quickly realised the need for a more efficient means of transporting coal from the pits of south-west Durham to the River Tees, and thence to markets at home and abroad, than the slow trains of packhorses that were the rule in the first quarter of the nineteenth century. That visionary was Edward Pease (1767–1858), an influential Darlington Quaker, merchant, philanthropist, coal owner and woollen manufacturer.[1]

During these eventful years two dominant issues were to have profound implications for Darlington – the choice between a canal or a railway, and the question of the route. Edward Pease would be well aware that as early as 9 November 1767, at a meeting held at the Posthouse Inn, the leading merchants of Darlington and surrounding districts had commissioned a survey for a canal from the Auckland coalfield to navigable water on the River Tees at Stockton.[2] James Brindley, the foremost canal engineer of the day, was approached, and in the summer of 1768 his son-in-law Robert Whitworth, assisted by George Dixon, began work on the survey. Whitworth's Report was presented in October of the following year, and recommended a route from Winston on the River Tees, 4 miles downstream of Barnard Castle, by way of Cockerton, a village on the north-west outskirts of Darlington, to Stockton-on-Tees, with branches to Yarm, Croft and Piercebridge. The total length was 33½ miles and the estimated cost of construction £64,000. Brindley subsequently confirmed the feasibility of Whitworth's route, and their joint Report was considered by the Darlington promoters in July 1769. Because of an unfavourable financial climate nothing further was done and the plans were quietly shelved.[3]

Towards the end of the eighteenth century, however, it became clear that Stockton's maritime functions were gravely impeded by the unsuitability of the Tees as a navigable river. In particular, the stretch downstream to the mouth was excessively meandering and prone to silting, with the result that large sea-going vessels over 150 tons had to be loaded and unloaded at the small out-ports of Newport and Cargo Fleet further downstream. To remedy the situation the Tees Navigation Company was formed in 1805 as the sponsoring body for river improvements. Its first major achievement was the completion of the 220-yard Mandale Cut, which shortened the distance between Stockton and the river mouth by 2½ miles. It was at a dinner in the Town Hall to celebrate the opening of the cut, on 18 September 1810, that Leonard Raisbeck, Recorder of Stockton and Solicitor to the navigation company, in the chair, seconded by Benjamin Flounders of Yarm, proposed a successful motion for the appointment of a committee 'to enquire into the practicability and advantage of a *Railway* or *Canal* from Stockton and Darlington to Winston, for the more easy and

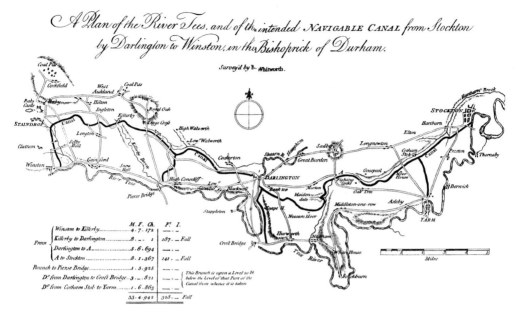

Plan of a proposed canal from Stockton by Darlington to Winston, which was surveyed by Robert Whitworth and published in 1772.

expeditious carriage of coals, lead etc.'[4] (As Maurice Kirby has pointed out, there was no precedent in the north-east for the construction of such a railway line over an extended distance,[5] and for some years after this meeting the 'Stockton Party' favoured a canal rather than a railway.)

The committee's Report was presented to a meeting held at the King's Head coaching inn, Darlington, on 17 January 1812, under the chairmanship of George Allen. It concluded: 'Either a canal or a railway would not only be productive of considerable advantage to the country in general, but would likewise afford an ample return to the subscribers.' The meeting accordingly resolved: 'That a survey of the Country or District through which the proposed Canal or Railway is intended to pass be forthwith made by Mr John Rennie. That he be instructed to make a report as to the practicality of those several measures, their comparative advantages, and the best line or course or extent of each; with an estimate of the expense which will attend the carrying of it into effect.'

The engineer and canal-builder John Rennie (1794–1874) presented his Report on 13 August 1813 and found in favour of a *canal* to the River Tees by way of Cockerton. (The soubriquet 'Cockerton Docks' was no mere jest, as it later became, since at this time they were a real possibility. In the light of subsequent developments it is ironic that the narrow packhorse bridge across the beck at Cockerton proved a long-standing obstruction to the expansion of the Darlington tramway system beyond the borough boundary here.)

Rennie effectively confirmed the validity of Whitworth's route of 1768, but this time at an inflated estimated cost of £205,618. A year and a half elapsed before Rennie's Report and Estimates were considered by the committee, on 7 February 1815. Because of the adverse economic climate of the times, the cost proved to be practically prohibitive, his proposals were quietly shelved, and nothing of any consequence was done for some three years.[8] Of

> ## JOHN RENNIE 1761–1821
>
> It was as a distinguished canal engineer that in 1813 John Rennie was engaged to make a survey of an intended route between the south-west Durham coalfield and the coast. Not surprisingly, he found in favour of a canal, on the same route which had been proposed by Whitworth and Brindley fifty years earlier. He must have been piqued, therefore, when the same committee in 1818, having rejected his advice, called upon him to carry out a further survey, but this time for a *railway*, in collaboration with the civil engineer Robert Stevenson (*q.v.*). Rennie refused the invitation, ostensibly on the grounds of professional jealousy,[6] but in reality because of the proposed collaboration with Stevenson. In a subsequent letter to the committee dated 26 December 1818 he wrote: 'I have been accustomed to think for myself in the numerous Publick Works in which I have been engaged, many of them of infinitely greater magnitude and importance than the Darlington Railway. If the subscribers to this scheme have not sufficient confidence in me to be guided by my advice I must decline all further concern with it.' The letter marked the end of Rennie's connection with the Stockton and Darlington scheme, 'and he did not live to learn how wrong he had been to belittle its importance'.[7]

course, Rennie's proposals had gratified the party that favoured a route serving Darlington, and eventually spurred the party that favoured a more direct canal route to Stockton to further action. One of their number, Christopher Tennant, a merchant of Stockton with interests in the lime trade at Thickley, near Shildon, financed another survey to that end, which was carried out by George Leather of Bradford. He duly recommended the more direct route, by-passing Yarm and Darlington and crossing the River Skerne at Bradbury. At a crowded public meeting in Stockton Town Hall on 31 July 1818 Tennant's scheme for the Stockton and Auckland canal was received with acclamation, despite its estimated cost of over £200,000 and a vote was taken to proceed with it. (Only Leonard Raisbeck spoke against the canal. Significantly, Edward Pease from Darlington and Benjamin Flounders of Yarm, supporters of Raisbeck, were also present at this meeting, although they had no vote.) In November Tennant led a deputation to London to raise the necessary four-fifths of the cost of construction so that Parliamentary approval could be obtained.[9]

The Darlington party were understandably dismayed, and in turn responded by 'reaching for their old blunderbuss, the canal scheme by way of Darlington, and discharged it repeatedly, though with lessening effect'. Then Providence intervened and put a new weapon into their hands. Yarm, an ancient borough on the south bank of the River Tees, had lost its pre-eminence as a port to the borough of Stockton, 7 miles downstream. Not surprisingly, it had led the way in organising opposition to the more direct, northern scheme. Several of the more zealous supporters of the original canal scheme lived in Yarm, including the Quaker Richard Miles, a lead agent with interests in the slate and timber trades; Thomas Meynell, a landowner; and his steward, Jeremiah Cairns. The latter was related by marriage to George Overton of Lanthelly near Brecon, an engineer with considerable experience in laying down tramways in the colliery districts of South Wales. Cairns wrote to him concerning the proposed canal, which initiated a long series of letters in which Overton 'pointed out with much cogency the superiority of a tram-road, many miles of which he had used and laid down'. Cairns showed these letters to Richard Miles, who, together with Leonard Raisbeck, had already been approached by Jonathan Backhouse (1779–1842), senior partner in Backhouses Bank and a leading Darlington instigator of the more direct route. Miles seems to have subsequently played a more important role in the promotion of railways than he has been given credit for. According to Cairns, 'he saw in the establishment of an iron road the solution of our difficulties'. Edward Pease said of him 'he daily appreciated the value and importance of railways, and he was among the first in the kingdom who gave them consideration as objects of public utility'.

A committee was duly appointed, which met on 4 September 1818, under Thomas Meynell's chairmanship, 'to consider the propriety of opening up a line of communication between Stockton and the collieries'. The main topic of discussion was again the relative advantages of a canal or a railway, and the probable route. A substantial step forward was taken when Leonard Raisbeck moved a successful resolution: 'estimates to be made for a line from Stockton by way of Darlington to the collieries, with branches to Yarm, Croft and Piercebridge, as well by the joint mode of having a canal and railway [as favoured by a group led by Jonathan Backhouse] as by railway the whole distance'[10] [as favoured by a group dominated by Thomas Meynell, the Darlington builder Thomas Botcherby, Richard Miles, and Edward Pease]. To implement that resolution it was also resolved to commission George Overton to undertake a survey – his brief, to find the most suitable route for a *railway* or a canal which (of course!) would serve both Darlington and Yarm.

In fact, Overton had anticipated that commission, since Meynell had written to Richard Miles, the committee Secretary, suggesting that he should issue the invitation privately because 'some of the Gents feel a delicacy in appearing to employ any Engineer but Mr Rennie'. As a result Overton came north, somewhat clandestinely, in advance of the committee meeting on 4 September, bringing with him, as his assistant surveyor, David Davies of Llangattock, Crickhowell, and was able to walk over the terrain the previous day, 3 September! He completed his survey, in the remarkably short time of seventeen days, on 20 September. In his Report to the committee, while he demonstrated that a canal from Stockton to Darlington was feasible, he recommended a 'tramway' traversing the whole route from Etherley Colliery by way of Darlington at an estimated cost of £124,000. That part of Overton's Report is truly historic because it introduced for the first time the concept of a 'Stockton and Darlington Railway' as we know it today, and established a route which (with significant modifications) it was to follow in practice. In fact, the title was officially applied to the scheme the following day, 21 September, and a railway committee was established.

THE FIRST S&DR ACT

In the meantime, the committee had authorised the necessary Parliamentary notice, issued on 9 September. In order to comply with the rules of Parliament, it was essential that a further application for an Act, which would incorporate the undertaking as a Joint Stock Company and sanction the construction of either a canal or a railway, should be made before 30 September of that year, 1818. Another important precondition was that plans of undertakings that required authorisation by Act of Parliament – for example, railways, canals, turnpike roads, docks, harbours and piers – were to be deposited and made available for public inspection at the time of the Parliamentary Bill needed to authorise the undertaking, according to the Standing Orders of the House of Commons 1792, confirmed by an Act of 1837 (QDP (M)). All Railway Acts therefore included in the preamble the

> ### ROBERT STEVENSON 1772–1850
>
> The grandfather of Robert Louis Stevenson, this Robert Stevenson was born in Glasgow, where he first went to work for his stepfather, Thomas Smith, whom he succeeded as engineer to the Northern Lighthouse Board – a post he held for fifty years. He designed and constructed twenty lighthouses, the most important being the 100ft high Bell Rock Tower, built in 1812 on a dangerous reef submerged at every high tide in the midst of the busy fairway of ships making for the Forth and Tay estuaries. Stevenson was the inventor of intermittent flashing lights for lighthouses, and a system of alternating red and white lights described by Sir Walter Scott as 'that ruddy gem of changeful light'. Although he must have seemed an odd choice to advise on Overton's route for a railway, in fact his practice covered the whole field of general engineering. He designed many bridges, including a novel form of suspension bridge. George Stephenson acknowledged that it was from hints in Stevenson's *Report on Railways* that malleable iron rails (*q.v.*) were developed. When Walter Scott undertook a celebrated voyage round the Firth of Forth he wrote: 'The Captain in Chief of this Empire is Mr Stevenson the surveyor – a most gentlemanlike and modest man, and well-known for his scientific skill.'

customary formula '. . . and whereas a Survey has been taken of the Lines of the said Railways or Tramroads, and a Map or Plan with a Book of Reference thereto, describing the same, and deposited with the Clerk of the Peace for the County of Durham . . . remains with the Clerk of the Peace, so that all Persons may examine and make extracts, or copies of the same'. George Overton's hastily prepared Plan and Sections (hereafter referred to as Overton (1)) were therefore deposited with the Clerk of the Peace at Durham on the very day of the deadline, 30 September. This, the first of Overton's three surveys, is significantly entitled 'Tramroad from West Auckland Collieries to Stockton via Darlington'. The course of Overton's route is indicated by numbered plots of land through which the line is drawn, for ease of identification.[11]

It appears that the committee had second thoughts about the merits of Overton's line, because on 25 September 1818 they proposed that Rennie be re-engaged to pronounce judgement on it. At the same time, an invitation was extended to Robert Stevenson, engineer of the Bell Rock lighthouse and several railway projects in Scotland, to participate in any new survey. In the event, Rennie turned down the invitation. On the other hand, Robert Stevenson did proceed with a survey, but it was completed too late to have any real bearing on events.[12] In the absence of any detailed critique by Rennie or Stevenson, therefore, Overton's original Report and recommendations were presented to a meeting of potential subscribers held in Darlington Town Hall on 13 November 1918, together with a new pronouncement which had been obtained from Rennie, setting out the following principles:

> First . . . a Canal will convey goods with the same facility upwards as downwards, whereas a Railway conveys them with more facility downwards than upwards: secondly, when the descending and the ascending trade is equal in quantity, a canal is preferable to a railway; but, where the descending trade very much *exceeds* the ascending, a Railway is preferable to a Canal.

The principal speakers at the meeting were Jonathan Backhouse, John Grimshaw and Edward Pease. The latter's argument was that there was a certain return of 5 per cent on a capital of £120,000, basing his estimates on the tolls from the Coal Road from Darlington to West Auckland. His reasoning carried the day, and it was decided to proceed with a second application to Parliament for an Act that would authorise the building of a *railway* throughout. (In the event, the word adopted in the Act was a *tramway*, 'on the plan and estimates given by Mr Overton'.) This meeting was a truly seminal one since the canal or railway controversy was finally resolved.

Events were moving fast. As a result of decisions taken at the meeting on 13 November, a prospectus was issued in December 1818, drawn up by Joseph Pease (1799–1872), the second son of Edward Pease, who was only nineteen years of age at the time. It was entitled: 'Prospectus for making a public railway from the collieries near Auckland to Darlington, Yarm and Stockton, for the supply of the South and East parts of the County of Durham and the North Riding of Yorkshire with coals, and for the general conveyance of merchandise. To be established by subscription, in shares of one hundred pounds each . . .'. It thus laid considerable stress on the potential contribution of the line to the development of internal trade:

> Local circumstances entitle this undertaking to the support of the general Public. The line passes through a populous district in which an extensive Trade already exists (this maintains Trade and Capital in their wonted channel without detriment to old establishments). General merchandise will be carried from the coast *upwards* and coal, lime, lead, Bluestone for repair of Roads and agricultural purposes from the interior of the county *downwards*. Coal and lime abound in the western parts of the county of Durham, from which the South East and North Riding of Yorkshire are supplied. By the facilities thus given the carriage will be reduced nearly *one-half* the present charge and that over a district containing not less than 40,000 inhabitants.[13]

The estimated cost of the line had been reduced to £113,000. Within a week, £25,000 had been subscribed, without exertion by the promoters,[14] the largest contributors being Edward Pease, Joseph Pease, Jonathan Backhouse, Joseph Gurney, all Quakers of Darlington, Thomas Meynell and Jeremiah Cairns of Yarm, William Chaytor, Thomas Richardson, Benjamin Flounders, John Kitching and William Kitching Junior. On 24 December, the day the subscription list was closed, Joseph Pease declared to Quaker 'cousins' in Leeds that the scheme for a 'Darlington Railway' was 'so popular among Friends that about £80,000 stands in the names of our Society'.[15]

The first S&DR Bill was narrowly defeated by 106 votes to 93 on 5 April 1819. Two reasons have been advanced for its rejection. First, it was found that owing to the haste with which the plans were prepared and deposited, they were incomplete. Second, according to Francis Mewburn (1785–1867), the company Solicitor, as a result of the lobbying of the two most influential landowners in the district, the Earl of Eldon and the Earl of Darlington (later the first Earl of Cleveland). They strongly objected to Overton's line because it cut deeply into their estates. While the Earl of Eldon was eventually bought off, Lord Darlington 'remained implacable in his opposition to a scheme which in his view was "harsh and oppressive to the interests of the country through which it is intended the railway should pass".'[16] His real concern, however, almost to the point of obsession, was the damage that the railway would inflict on his fox coverts![17] (Francis Mewburn was to write in his diary much later: 'The difficulties, pain and anguish which I endured during my sojourn in London while soliciting the Bill can scarcely be imagined.')[18] To meet these objections, on 9 July 1819 the promoters gladly accepted George Overton's offer to conduct a second survey.

The promoters were not discouraged by their experiences, and ascribed their lack of success to 'a want of money, time and talent'.[19] 'Edward Pease never knew what it was to be beaten or feel cowed. The result was inspiriting to the Quakers, for they saw their strength.'[20] The promoters having learned their lesson, the Bill was re-presented in amended form in July 1819. In the meantime, Lord Eldon had accepted the compulsory purchase of land while Lord Darlington was placated when Overton succeeded in drawing a new route that avoided his estates. His proposals for a line of 37 miles, including branches – 9 miles shorter than Overton (1) – were deposited with the Clerk of the Peace for Durham County on 30 September 1819, under the title 'Plan and Section of the Intended Rail Way or Tram Road from Stockton by Darlington to the Collieries near West Auckland with several intended Branches of Rail Way or Tram Road in the County of Durham 1819 From the Line marked out by George Overton Engineer Dd Davies Surveyor' (Overton (2)).[21] The incorporation of the alternatives 'Rail Way' and 'Tram Road' calls for comment. It was not that the proprietors had a clear idea at this stage of the important distinction between the two, and were 'keeping their options open'. Rather, 'Rail Way' and 'Tram Road' were used interchangeably in a general sense to indicate an 'iron way' as opposed, for example, to a turnpike road or a canal.

This time the Bill was delayed by the death of King George III on 29 January 1820, which brought about a Dissolution of Parliament and a general election. A plaque on the wall of the George and Dragon Hotel in the High Street, Yarm, commemorates a meeting of the committee on 12 February 1820 at which it was decided to postpone operations for a year until the following Parliamentary session of 1820–1. The wording of the plaque rather disingenuously accords to the meeting rather more importance than it warranted: 'S&DR. In the Commercial room of this hotel on the 12th February 1820 was held the Promoters Meeting of the Stockton and Darlington Railway – the first public railway in the world. Thomas Meynell, Esq., of Yarm, presided.' What the plaque does establish is the influence of the 'Yarm Party' at this time.

The delay in the progress of the Bill provided the promoters with an opportunity to approach Overton yet again to come forward with a revised and still shorter route. Understandably, this caused Overton some frustration, combined with irritation at the fact that the committee had again sought the advice of Robert Stevenson. However, after some negotiation, Overton was ultimately prevailed upon to undertake a fresh survey, which he completed by the end of August 1820. His Plan and Sections (Overton (3)) were deposited with the Clerk of the Peace at Durham on 30 September 1820 under a title identical to that of Overton (2), which had been deposited precisely one year earlier. It was this third plan that formed the basis of a renewed submission to Parliament.[22]

When the delayed and amended Bill was lodged with the Private Bills Office of the House of Commons the estimated cost had been reduced to £92,000. By the end of March 1821 it had passed the committee stage, 'in high stile', according to Leonard Raisbeck,[23] and with only one hostile petition presented, it became the First S&DR Act when it received the Royal Assent on 19 April 1821 (1 and 2 Geo. IV, cap. 44). It was entitled 'An Act for making and maintaining a Railway or Tramroad from the River Tees at Stockton to Witton Park Colliery, with several branches therefrom, all in the County of Durham'. In fact, it was not the first Railway Act, but the twenty-first to be passed by Parliament since the beginning of the century, and the draft by Francis Mewburn was largely modelled on the Berwick–Kelso Railway Act of 1811.[24] The provisions of the Act extended to sixty-seven pages and incorporated the whole of the law relating to railways up to that time. It authorised 'the making and maintaining a Railway [sic] or Tramroad from the River Tees at Stockton, west to Darlington then north-west to Witton Park Colliery [near the south bank of the River Wear] with several branches therefrom, all in the County of Durham'. The designation of Stockton first in this document is probably the reason why this railway has been enshrined in history as the 'Stockton and Darlington', and not the reverse. J.S. Jeans maintained that 'It was mainly for the purpose of conciliating Stockton – a kind of sop to Cerberus – that the name of that town was allowed priority in the designation of the new line, and not because Stockton had contributed resources towards the undertaking entitling her to that distinction.' Although Jeans is clearly prejudiced in Darlington's favour, there may well have been an element of truth in his assertion.

The Act authorised the raising of £100,000 of capital in the proportion of £80,000 in shares and £20,000 by loan. Of the seventy-nine subscribers listed in the Act the following, for a variety of reasons, feature prominently in this account: Joseph Pease the Younger, Thomas Benson Pease, Edward Pease, Jonathan Backhouse the Younger, John Backhouse, Daniel Milford Peacock, clerk (that is, clergyman), Leonard Raisbeck, Lord Viscount Barrington, Robert Botcherby, Joseph Gurney, Joseph John Gurney, John Kitching, Thomas Meynell, Richard Pickersgill, Francis Storey and William Tate. In the words of the Act the subscribers 'are hereby instituted into a Company for making and maintaining the Railways or Tramroads and shall for that purpose be one Body Corporate, by the name and style of "The Stockton and Darlington Railway Company", and by that name shall have a perpetual succession and common seal'.[25]

FINANCE AND THE QUAKER CONNECTION

In the 1821 session the Bill almost came to grief because of non-compliance with the standing order that four-fifths of the amount of share capital should be subscribed before the Bill went into committee. Francis Mewburn, who was the Parliamentary Agent, found that they were £7,000 short, and he was personally unable to raise this sum in London. With only days to spare, he immediately wrote to Leonard Raisbeck, with whom he shared the duties of Railway Solicitor, in Stockton, who informed Edward Pease of the crisis. Characteristically, Edward Pease promptly subscribed the entire £7,000 himself,[26] in addition to his original subscription of £3,000. 'It was this gesture of confidence in the enterprise at a critical moment which marked the ascendancy of the Pease family over the conduct of the S&DR. Henceforth it was known as "the Quaker line",[27] much to the disgust of William Chaytor who had earlier resigned as chairman of the management committee in protest at the growing dominance exercised by the Pease–Backhouse alliance.'[28]

This was not the first time that Pease had saved the day financially. At the meeting held at Yarm on 12 February 1820 he jotted down notes of what had already become a crisis meeting of the committee:

> When Mr Raisbeck said 'Gentlemen your subscription deed is £10,000 short of the amount required, the mail is nearly due and if I do not go up by it the bill cannot be lodged this session' the committee looked at each other paralysed. Then Mr E Pease said 'I did not intend to subscribe more. I can at present spare £10,000, and I do not incline to loose our Bill. I will sign for £10,000, with a three years option to take bonds of £100 shares at par.'[29]

These quotations underline the leading role of the Quaker Edward Pease as financier to the S&DR in its infancy. What is not so readily apparent is the crucial role in the financial affairs of the S&DR that was played by the Quaker community as a whole – the 'family' of the Society of Friends. There was an especially strong concentration of Quaker families in Darlington of whom the Peases and the interrelated Backhouses (Bankers) were most prominent. Darlington was familiarly known as 'The Quaker Town', in the same way as the S&DR itself was known as 'The Quaker Line'.

In addition, however, the Peases were allied through intermarriage to other wealthy Quaker families elsewhere, notably the Gurneys of Norwich and London. For example, on 20 March 1826 Edward Pease's second son Joseph married Emma, daughter and co-heiress of Joseph Gurney of Norwich. Edward Pease's cousin Thomas Richardson was the founder of the influential discount house of Overend, Gurney and Company of Lombard Street in the City of London. Richardson's partner in this enterprise was Samuel Gurney, the nephew of Joseph Gurney. When the S&DR was first promoted Richardson took up £5,000 worth of stock and thus became the largest investor at that time. Later, when the cost of the works exceeded George Stephenson's estimate, Messrs Overend, Gurney made a loan of £20,000 to the company pending a new issue of stock. Because of these close ties between the extended Quaker 'family' in Darlington and Quaker communities in other parts of the kingdom, they could call upon financial resources in addition to their own when the need arose, as indeed it did throughout the independent existence of the S&DR and its associated companies. Moreover, such merchant and banking families as the Gurneys were powerful 'political' as well as financial allies and did much to smooth the passage of successive S&DR Bills through Parliament. It has been said that without these powerful connections the pioneer venture might have foundered.

The point is well put by Maurice Kirby, who writes, for example, of the S&DR's record of capital formation:

> In this respect the company was unique in the history of UK railway enterprise as a public joint stock concern which was, in effect, a close family partnership. Throughout its existence as an independent entity the company relied heavily upon the Society of Friends for its capital and borrowing requirements. This had been a notable feature of the enterprise at the time of its inauguration in 1821, but by the mid-1840s the pattern of share ownership had come to be concentrated in the hands of the Quaker Pease family of Darlington which by virtue of its extensive intermarriage with a number of Britain's leading Quaker business dynasties was able to tap a private, nationwide capital market. . . . The ascendancy of the Peases and their immediate Quaker associates was assured as they became major traffic senders on the S&DR network following their own extensive investment in local mineral development. . . . It would be true to say that the expansion of the Stockton and Darlington Company was never at any time constrained by lack of financial resources. It is impossible, therefore, to examine the capital structure and investment strategy of the company without taking account of kinship ties of unusual strength and geographical dispersion.[30]

The financial involvement of the Pease family in the affairs of the S&DR contributed to their dominance of its policy-making and management structures. Unlike succeeding railway enterprises that relied upon a class of professional managers, the S&DR depended to a large extent on the accumulated experience and business acumen of this one dynasty. A principal strategy of the S&DR was to promote nominally independent companies. In reality, there was a substantial degree of overlap between the directorates of the S&DR and these associated companies, with a central core drawn from the Pease family. 'This ensured continuity of decision-making throughout the S&DR's independent existence . . . The Peases consistently revealed themselves as dynamic, forceful and innovative.'[31]

In the course of time, the word 'proprietor' came to signify the most influential owners of the company. However, it originated in a much more general use of the word which derived from the first S&DR Act of 1821, where subscribers are defined as 'proprietors of every such share [in the Company]'. The first share certificates repeated the wording of the Act, as in the earliest extant example:[32] 'Stockton and Darlington Railway. Incorporated by an Act of Parliament Passed in the Second Year of the Reign of His Majesty George IVth. These are to certify that Thomas Newman . . . is *Proprietor* of the Share No 535 of the Stockton and Darlington Railway and . . . is entitled to the Profit and Advantage of such Share. Given under the Common Seal of the Company the 25th day of July in the Year 1823.' The certificate is signed by Leonard Raisbeck and Francis Mewburn as 'Chief Clerks to the Company'. (They were also joint Solicitors.) Hereafter, *proprietor* is used in preference to *director*, although before long the two words became synonymous.[33]

EDWARD PEASE AND GEORGE STEPHENSON

By this time, Edward Pease was taking a considerable personal interest in the project. From the beginning he had favoured a railway as opposed to a canal for the whole distance from the collieries to Stockton. One of his principal objectives – that the route should be by way of Darlington, where his woollen mills and other interests would benefit from the reduced costs of transport – was now assured. It was at this point that George Stephenson (1781–1848) became involved in the enterprise. The 'Father of Railways' was born on 9 June 1781 in a cottage, rather grandly called 'High Street House', by the side of the wagonway leading from Wylam Colliery near the River Tyne, 10 miles upstream of Newcastle. His father Robert was the fireman of the pumping engine at the colliery at the time. George's first mining job, at the tender age of thirteen, was as corfe-bitter at nearby Dewley Burn Colliery – picking out stones – 'bats and brasses' – from the newly won coal. From that he graduated through a number of increasingly responsible posts at various collieries in the neighbourhood – supervising a horse-drawn gin, assistant fireman (to his father), fireman, engineman, brakesman. It was while he was employed as brakesman at Ballast Hill, Willington, downstream of Newcastle, that he married his first wife,

George Stephenson by Henry Perronet Briggs (c. 1791/3–1844). (National Railway Museum/Science & Society)

Fanny Henderson of Newburn, on 28 October 1802. The young couple went to live in a cottage at Willington Quay nearby, and it was here that George Stephenson's only son Robert (1803–59) was born on 16 October 1803. In 1805 Stephenson moved to take up the post of brakesman at West Moor Colliery, Killingworth, 4 miles north of Newcastle upon Tyne. Sadly, his wife died there in the following year shortly after giving birth to a daughter. Devastated at this loss, Stephenson took up employment for a brief while in Scotland as engineman at Montrose Spinning Mills before returning to his old job at West Moor. In 1812 he was appointed engine-wright at the High Pit Colliery, Killingworth, which the Grand Allies had sunk the previous year.[34] In addition, he was put in charge of all the machinery in the collieries of the Grand Allies at a salary of £100 p.a.

It is clear that George Stephenson had already made a name for himself on Tyneside as the inventive engine-wright of Killingworth, with unrivalled experience in the laying out of colliery railways and the construction of locomotives, when he came to the notice of Edward Pease. According to Francis Mewburn – in an entry in his diary long afterwards, in 1862 – '[the success of the S&DR] resulted from the ability of Edward Pease to recognise the character and the genius of a Northumbrian colliery engineer whose work was as yet hardly known beyond his native county – George Stephenson'.

A truly historic meeting between Edward Pease and George Stephenson took place in Darlington on 19 April 1821. There are four versions of the circumstances that led up to this encounter, and what then took place. According to Samuel Smiles' rather colourful account, it was George Stephenson who took the initiative by walking with a companion, Nicholas Wood, the viewer at Killingworth Colliery, without any invitation, from Killingworth to Edward Pease's home at what is now 146 Northgate, Darlington.[35] Opposite the house there remains to this day a boulder, a 'glacial erratic', called the Bulmer's Stone after a town crier of that name who used to mount it to deliver a summary of the news received by post-chaise from London. 'In a memoir of Francis Mewburn, published in 1867, there is an account of the first meeting between George Stephenson and Edward Pease: "At the behest of Pease, old George with Nicholas Wood barefoot walked to Darlington, shoeing themselves near Bulmer's Stone. Neither was ever backward in admitting this, for neither was ashamed, for each had the true stuff of men."' (There is a splendid mural depicting this incident in the children's section of the Public Library in Crown Street.)[36]

The quotation comes from *The Diaries of Edward Pease*, edited by his great-grandson Sir Alfred Pease. The second version of the meeting between Edward Pease and George Stephenson is given in that edition, and lends credibility to Smiles' account:

One day in 1821 Edward Pease was writing in his room when a servant announced that two strange men wished to speak to him. He was busy and sent a message that he was much too occupied to see them. The door had no sooner closed than he laid down his pen and wondered whether he had done right; he then rose from his chair and went downstairs. He asked where they were, and was told they were in the kitchen. Going into the kitchen he found them, and they gave their names as Nicholas Wood, viewer at Killingworth Colliery, and George Stephenson, engine-wright at the pits. Mr. Pease sat down at the end of the kitchen table to learn their errand. Stephenson handed him a letter from Mr. Lambert, the manager of Killingworth, recommending Stephenson to the notice of Mr. Pease as a man who understood laying down railways.[37]

The third and more credible version is rather more mundane. It is contained in a first-hand account that Nicholas Wood himself gave in the course of a speech he delivered in 1862 in memory of George and Robert Stephenson. According to Wood, Edward Pease himself took the initiative and invited George Stephenson to meet him at his home by appointment, in a letter delivered in person by John Dixon.

> The incident is given by Smiles not quite correctly. The fact is we rode on horseback from Killingworth to Newcastle, a distance of five miles, travelled thence by coach, thirty-two miles, to Stockton, then walked along the proposed line of railway [Overton's route], twelve miles, from Stockton to Darlington. We had then the interview with Mr. Pease, by appointment, and afterwards walked eighteen long miles to Durham, within three miles of which I broke down, but was obliged to proceed, the beds being all engaged at the 'Traveller's Rest'.

According to Smiles, the couple missed the last coach because they became engrossed in their conversation with Edward Pease. In any event, they had not booked places on the coach but intended to return to Newcastle 'by nip', that is by tipping the coachman.

A fourth version is given by Stephenson himself in a document he wrote just before he died. *A Short Account of the Commencement of the Stockton and Darlington Railway* is dated 20 April 1848.[38] It was prompted by a letter which he had seen in a newspaper praising Edward Pease as a 'neglected man of science' who 'first conceived the idea, and so successfully carried out the first railway for the conveyance of goods and passengers'. Stephenson wrote to Nicholas Wood enclosing this description and asked him to write to the paper to 'shew that Edward Pease is not the man of science who has done so much for the country'. Because this account differs significantly from the version given by Wood, in particular as to whether the initiative for the meeting came from Pease or from Stephenson himself, it is worth quoting the relevant section in full:

> Mr Stephenson having succeeded so well with his Locomotive engine on the Killingworth, Springwell and Hetton railways which had been constructed under his care, he was advised by many of his friends to apply to the Stockton and Darlington Company to be appointed their engineer for the construction of the line for Locomotive engines – Mr Stephenson having obtained several letters of recommendation from his friends and accompanied by Mr Nicholas Wood started from Newcastle to Stockton by Coach with the intention of seeing Mr Raisbeck to whom he had letters of recommendation. Upon arriving at Stockton in the evening and finding that Mr Raisbeck was not at home they then proceeded to walk to Darlington after dusk in a very stormy night of snow and arrived at that town at a very late hour and took up their quarters at a small Inn opposite the church – The next morning they called upon Mr Edward Pease and Mr Jon'n Backhouse. Mr Stephenson having given his letters of recommendation to these gentlemen, a short conversation took place in which Mr Stephenson explained the powers of his Locomotive engine and recommending the substitution of a railroad in the place of a tram road. After having heard Mr Stephenson's statement Messrs Pease and Backhouse said a deputation from the Darlington Co should be sent to Killingworth to inspect the engines.

Whatever the circumstances, it is clear that George Stephenson made a very favourable impression on Edward Pease for, shortly after, the shrewd Quaker said: 'There was such an honest, sensible look about George Stephenson, and he seemed so modest and unpretending, and he spoke in the strong Northumbrian dialect of his district, and described himself as "only the engine-wright at Killingworth; that's what he was".'[39] This first meeting of the two men proved to be the beginning of a lifelong friendship. In a curious reversal of the usual relationship between patron and protégé, many years later it was Stephenson who presented Pease with a 'handsome gold watch – which he was proud to exhibit – bearing these words: "Esteem and gratitude: from George Stephenson to Edward Pease"'.

The day after the meeting Pease wrote to Stephenson to inform him of the passage of the Bill 'for the Darlington Railway', and making informal enquiry as to whether he would be willing to act as surveyor and engineer for the construction of the line. Stephenson replied from Killingworth on 28 April. The letter is instructive for the light it sheds on George Stephenson's honest and straightforward method of doing business, and his forthright character, which had no doubt commended him to Edward Pease:

> From the nature of my engagements here and in the neighbourhood, I could not devote the whole of my time to your Rail Way, but I am willing to undertake to survey and mark out the best line of Way within the limits prescribed by the Act of Parliament and also, to assist the Committee with plans and estimates, and in letting to the different contractors such work as they might judge it advisable to do by Contract and also to superintend the execution of the work. And I am induced to recommend the whole being done by Contract under the superintendance of competent persons appointed by the Committee.
>
> Were I to contract for the whole line of road it would be necessary for me to do so at an advanced price upon the Sub Contractors, and it would also be necessary for the Committee to have some person to superintend my undertaking. This would be attended with an extra expense and the Committee would derive no advantage to compensate for it. If you wish it, I will wait upon you at Darlington at an early opportunity when I can enter into more particulars as to remuneration, etc. etc.–[40]

We know from an account of changes that Stephenson later submitted that he did in fact 'wait upon' Pease again on 22 May, and that he spent the following three days 'looking over the country' at a charge of two guineas a day.

The first meeting between the two men took place on the very day, 19 April 1821, that the first S&DR enabling Act received the Royal Assent. At this stage, horses were envisaged as the motive-power, and George Stephenson's first contribution had been to persuade Edward Pease that steam locomotives, hauling wagons on iron edge rails, would be much more efficient and economic than a horse-drawn plate-way or tramroad in the South Wales manner, as George Overton intended.

According to Alfred Edward Pease (in his edition of his great-grandfather's *Diaries*), even before this historic meeting Edward Pease 'had long satisfied himself as to the soundness of his [George Stephenson's] idea "that a horse on an iron road would draw ten tons for one ton on a commonroad", and to use his own words, "I felt sure that before long the railway would become the King's Highway". Stephenson told him that the locomotive that he had made to run on the pit railway was worth fifty horses. 'Come over to Killingworth and see what my Blutcher [*sic*] can do – seeing is believing, Sir.' In the summer of 1821 Edward Pease and his cousin Thomas Richardson accepted the invitation and visited Killingworth colliery. There they saw George Stephenson's steam locomotive in action hauling a coal train on the wagonway that ran by the side of his cottage at West Moor. The party were even persuaded to mount the footplate. 'To say that Edward Pease was impressed by the experience is an understatement.' As he commented 'in truly visionary terms', in a letter to Thomas Richardson of 10 October 1821:[41]

Don't be surprised if I should tell thee there seems to us after careful examination no difficulty of laying a railroad from London to Edinburgh on which waggons would travel and take the mail at the rate of 20 miles per hour, when this is accomplished steam vessels may be laid aside! We went along a road upon one of these engines conveying about 50 tons at the rate of 7 or 8 miles per hour, and if the same power had been applied to speed which was applied to drag the waggons we should have gone 50 miles per hour – previous to seeing this locomotive engine I was at a loss to conceive how the engine could draw such a weight, without having a rack to work into the same or something like legs – but in this engine there is no such thing. . . . The more we see of Stephenson, the more we are pleased with him . . . he is altogether a self-taught genius . . . there is such a scale of sound ability.[42]

Presently, Edward Pease came to share the vision that George Stephenson expressed in the prophetic declaration: 'You will live to see the day when the railways will supersede almost all other methods of conveyance in the country – when mail coaches will go by railway and railroads will become the great highway for the king and all his subjects. The time is coming when it will be cheaper for a working man to travel on a railway than to walk on foot.'[43] Both Edward Pease and George Stephenson, therefore – but for different reasons – fully deserved the title 'Father of Railways' that was subsequently bestowed upon them.[44]

INCORPORATION

Much remained to be done before the railway could become operational. One of the many clauses of the First S&DR Act directed that 'The first General Meeting of the Company of Proprietors for putting the Act into execution, shall be held at Darlington Aforesaid, within six weeks of the passing of this Act', and that a committee should be elected to manage the affairs of the company. The first meeting of the shareholders of the newly formed company was accordingly held at the King's Head in Darlington on 12 May 1821. The committee's first task was to elect a new and smaller management committee, composed of fourteen of the principal subscribers. Thomas Meynell was reappointed as Chairman and Jonathan Backhouse as Treasurer. Edward Pease and his son Joseph were also elected onto a sub-committee of seven members, having the power to make contracts, as authorised in the Act, who were to be responsible for the detailed execution of the project. (It is significant that throughout his association with the S&DR Edward Pease himself was careful not to be elected to any official position on the management committee, although it is certain that because of his immense influence and authority he was the *éminence grise* behind most of its decisions.) Edward Pease and Thomas Meynell were authorised to approach various engineers regarding the construction of the line.

The committee duly met on 25 May and proceeded to adopt a design for a corporate seal for the Stockton and Darlington Railway. The Revd Peacock, Vicar of Great Stainton near Darlington, suggested the words PERICULUM PRIVATUM: UTILITAS PUBLICA – Private Risk in Public Service. It was a motto that perfectly expressed the business philosophy of the entire Pease dynasty. Significantly, the seal depicts three coal-carrying chauldron wagons, and part of a fourth, on rails, but they are hauled not by a steam engine, but a *horse*. At this stage, therefore, the company, although not Edward Pease himself, were not fully persuaded that steam locomotives would be the prime motive power for their enterprise. Behind the train of wagons is depicted a rural landscape and, in the distance, a winding-house powered by a static steam engine. (Not, as one might suppose, the pit-head gear of a colliery.)

Rather surprisingly, Edward Pease, even before the opening of the railway in whose promotion he had been so intimately involved, handed over his interest to his son Joseph. It was a wise move, for Joseph Pease was to be instrumental in overseeing the successful inauguration of the railway, and its subsequent operation as a viable concern.

The first Solicitor of the S&DR, who became, in consequence, the world's first railway Solicitor, was another Quaker, Francis Mewburn. As Solicitor, he was also Parliamentary Agent for the S&DR, preparing and presenting Bills that became Acts of Parliament, and was therefore instrumental in bringing the promoters' plans to fruition. At a celebratory luncheon in 1829, on the occasion of the opening of the Croft branch on 27 October, Mewburn even astonished the guests by 'venturing to prophesy that in a few years a railway would be made from Darlington to London, and so quick would be the travelling that passengers would leave the former place in the morning, arrive in London in time to go to the opera, and return home next day. This prophecy was received with shouts of laughter.'[45] In fact, Francis Mewburn was to further the progress of railways for forty years, from 1811 to 1860.

The S&DR seal and motto.

FRANCIS MEWBURN 1785–1867

Francis Mewburn, the S&DR's Solicitor, came to Darlington in May 1809. In 1811 he founded the Darlington Society for the Prosecution of Felons, and for two years ran a school for poor children in St Cuthbert's parish church. In 1831 he bought the mansion called Paradise near Coniscliffe Road, and promptly changed its name to Larchfield (after the lane which ran to the east of the grounds) 'because it was thought inappropriate for a solicitor to live in Paradise'! The Mewburn family motto was *Festina Lente* (Hasten with Caution), and it was his innate caution which provoked Edward Pease to say of him: 'Francis Mewburn, if I had no more courage than thee, I should attempt nothing at all, thou hast the heart of a chicken. I am determined to try out the railway.' Nevertheless, under Edward Pease's tutelage Mewburn's caution soon gave way to enthusiasm. Joseph Pease came to trust him, and he was not alone. In 1846 Bishop Maltby appointed Mewburn Chief Bailiff of Darlington, to act as his manager in the town. In the absence of a mayor, he was in effect the town's Chief Citizen, and when he died in 1867 the post died with him.[46]

THE SECOND S&DR ACT

The circumstances in which George Stephenson came to be appointed engineer to the S&DR are set out in full in the following chapter. It is sufficient to indicate here that in that capacity he was invited by the committee to undertake a fresh survey of Overton's line, upon which the First S&DR Act of 19 April 1821 was based, and to incorporate any deviations to that line which he considered desirable. The results of his labours were deposited with the Clerk of the Peace at Durham on 28 September 1822 under a title the first part of which is almost identical to that of Overton (3): 'Plan and Section of the intended Railway or Tramroad from Stockton by Darlington to the Collieries Near West Auckland in the County of Durham and of several Branches therefrom' (so far so good). It continues: 'and of the Variations and Alterations intended to be made therein respectively and of the Additional Branch of Railway Proposed and Made 1822 Geo Stephenson Engineer'.[47] The 'Additional Branch of Railway' refers to what later became known as the Croft branch. True to its title, Stephenson's plan includes the line of Overton's route, differently coloured, to indicate where his 'Deviations' begin and end. (These early railway Plans and Sections are truly impressive examples of the surveyors' skill. When unrolled, George Stephenson's plan measures 3 ft 6 in by 8 ft, while Overton's three submissions are only marginally smaller.)

Together with Leonard Raisbeck, Francis Mewburn was responsible for drafting the revised Bill, based on Stephenson's survey, which in due course became the Second S&DR Act when it received the Royal Assent on 23 May 1823. It was entitled 'An Act to enable the Stockton and Darlington Railway Company to vary and alter the line of their Railway, and also the line or lines of some of their Branches therefrom, and to make an additional Branch therefrom, and for altering and enlarging the Powers of the Act passed for making and maintaining the said Railway'. In addition to seeking authorisation for the deviations proposed from Overton's route, therefore, this Second Act incorporated changes and additions to the first which the company thought prudent in the light of experience, and largely on the advice of George Stephenson. Significantly, the First Act of 19 April 1821 made no specific mention of steam locomotives as motive power, but left that possibility open by providing that the railway should be operated 'with men or horses *or otherwise*'. Two clauses make the Act of 23 May 1823 quite unique. First, the company is given power to 'make, erect and set up one permanent or fixed Steam Engine at or near each of the inclined Planes' (referring to the Etherley and Brusselton inclines). Second, in the next section the word locomotive is used for the first time. The actual text is as follows:

> That it shall be lawful for the Company of Proprietors to make and erect such and so many *loco-motive* or moveable Engines, as the said Company shall from Time to Time think proper and expedient, and to use and employ the same in or upon the said Railways or Tramroads, or any of them, for the Purpose of facilitating the Transport, Conveyance, and Carriage of Goods, Merchandize and other Articles and Things upon and along the same Roads, and for the Conveyance of *Passengers* upon and along the same Roads.

(Francis Mewburn noted in his diary, apropos locomotives, that when he submitted the Bill to Lord Shaftesbury's Secretary, that gentleman 'could not conceive what it meant; he thought it was some strange and unheard of animal and he struck the clause out of this Act'. It was reinstated when Mr Brandling, the MP for Sunderland, and George Stephenson were sent to explain the matter to him: 'A very short study of the earliest history of the creature will show that the secretary's ignorance was quite excusable.')[48] We are therefore justified in according to the Stockton and Darlington Railway the title 'First in the World' since in this Enabling Act *locomotives* as motive power and the provision of a *passenger* service are for the first time given Parliamentary approval. Ironically, the full significance of the inclusion of these two words was by no means appreciated by the promoters of the S&DR at the time. In fact, the various committee minutes kept by the company contain no firm evidence of the change of policy from horse traction to mixed horse, fixed engine, and locomotive engine haulage, but there is much circumstantial evidence which suggests that it was George Stephenson who convinced the proprietors that these methods taken together were more efficient.

The Second S&DR Act of 1823 also permitted the company to reduce its capital from £82,000 to £74,300 in accordance with the reduced track mileage, and notwithstanding the inclusion of the additional Croft branch. It was undoubtedly the economies that were achieved as a result of Stephenson's Survey which largely persuaded Parliament to approve the Act. To be precise, whereas Overton had estimated the construction costs as £77,341 18s 8d, Stephenson's estimate was £60,987 13s 8d.[49]

Nevertheless, the building and operation of the railway was still a costly enterprise. The first and most pressing consideration now was to secure the backing of sufficient private subscribers to meet the balance of the cost not already raised. A new prospectus was drawn up and once again, among the most prominent subscribers, were members of the Pease family resident in Darlington, notably Edward and his son Joseph, and others of their Quaker friends and business partners. It was becoming patently clear why it had been dubbed 'The Quaker Line'. For example, the Treasurer, Jonathan Backhouse, acted as shares promoter, and on 9 August 1821 a Mr Thomas Peacock was appointed 'Collector of Monies for the S&DR'.

Of course, money in hand and safely lodged in the bank does not of itself build a railway. We must also consider the engineering works on which that money was spent and which, in turn, led to the inauguration of the S&DR on 27 September 1825.

CHAPTER 2
Foundations: Engineering

GEORGE STEPHENSON APPOINTED ENGINEER

The management committee, which met on 25 May 1821, 'desired Mr Meynell and Mr Pease, with a view to the construction of the railway, to ascertain the charges of engineers for undertakings of this kind'. Edward Pease, of course, favoured George Stephenson, and the outcome was that at a further General Meeting on 23 July 1821 it was resolved that 'in order to settle the line George Stephenson be imployed in the first instance to survey the one laid out by G. Overton',[1] with certain qualifications. (The sub-committee was authorised to write to Stephenson to obtain his charges for this work.) The qualifications were that he should state what deviations from Overton's line he thought desirable – that is, both inside and outside the authorised limits, 100 yards on either side of the approved Parliamentary line – and in the latter case to specify through whose lands the deviations would lead and to submit comparative estimates of their construction costs. (There seems to have been a tacit understanding that whoever was commissioned to undertake a survey would be appointed engineer to the line if it was adopted.) It is significant that the first resolution passed at this General Meeting read: 'That in the best information this Meeting has been able to collect a *Railway* be adopted as preferable to a *Tramway*, and that Land sufficient for a double Railway be purchased as soon as the precise line is definitely settled.' Implicit in the resolution is the reason why Overton's route needed to be re-surveyed. That was primarily designed for a horse-tramway or plate-way, whereas Stephenson's recommendation was for a 'RailWay' designed for steam locomotive traction on edge rails.[2] Heavily influenced by Edward Pease, the General Meeting took the crucial decision to back Stephenson's railway in preference to Overton's tramway. 'From the moment that George Stephenson met Edward Pease the name of George Overton was no more heard in Darlington. Presumably he returned to Wales a sad and disgruntled man.'

The decision to employ George Stephenson was considerably influenced by a letter of commendation dated 22 July 1821 from an eminent engineer, William James, of Chipping Norton, Oxfordshire, which was sent in advance to all the members of the General Meeting. After strongly urging the superiority of edge rails over flange rails, and of malleable-iron over cast-iron rails, James goes on: 'I feel it due to the Character and talents of the Northumbrians to declare that in the sciences of Mining and Mechanics we can in the South in no respect be compared with them . . . and with such Abilities as Mr. Stephenson's and Mr. Chapman's. . . . How sensible I am of the Value of these Gentlemen.'[3]

In conveying the sub-committee's invitation to Stephenson in a letter of 28 July, Edward Pease added: 'In making thy survey it must be borne in mind that this is for a great public way and to remain as long as any Coal in the district remains, its construction must be solid, and as little machinery introduced as possible; in fact, we wish thee to proceed in all thy levels, estimates and calculations with that care and economy which would influence thee if the whole work were thine own.' It is also clear that two matters for decision remained outstanding: 'it would be well to let comparative estimates be formed as to the expense of a double and single Railway and whether it would be needful to have it only double in some

JOHN DIXON 1797–1865

John Dixon was the grandson of the George Dixon who had assisted Whitworth in his canal survey of 1768.[4] When George Stephenson accepted the post of Surveyor to the line in 1821, he requested Edward Pease to 'send a suitable person along with me who knows the different Gentlemen's grounds through which we should pass'. A native of Darlington, John Dixon was the 'suitable person' nominated by the management committee. So began a long association with George Stephenson as one of his trainee 'assistant engineers'. When Stephenson was appointed engineer to the line, he made it clear that he could not devote his whole time to the S&DR. Two 'resident engineers' were therefore appointed under Stephenson's direction, one of whom was John Dixon and the other Thomas Storey. When the firm of Robert Stephenson and Company was founded in 1823 (*q.v.*), initially to build locomotives for the S&DR, a quite separate company known as George Stephenson and Son was formed to undertake railway surveys and construction. Among them were the Canterbury and Whitstable, the Liverpool and Birmingham, and the Grand Junction railways, for which John Dixon acted as survey engineer. As a servant of the company, John Dixon's own work, during this first part of his career in the 1830s, tended to be overshadowed by the reputation of his nominal employers. In 1842 he was appointed Chief Civil Engineer by the S&DR, soon extended to 'inspecting engineer' when his duties came to include the oversight of contracted-out locomotive building and the maintenance of the permanent way. It was only with this appointment – truly a homecoming – that John Dixon achieved a measure of independence and recognition in his own right. When he was later appointed engineer-in-chief of the S&DR he settled permanently in Darlington at the mansion Belle Vue. He continued to serve the company with distinction, and introduced 'cost analysis' programmes that did much to maximise the company's profits. John Dixon typified the dedicated professionalism of locally recruited senior personnel of the S&DR.

part and what parts, also comparative estimates as to the expense of Malliable or Cast Iron'. On 2 August George Stephenson replied from Killingworth, implicitly accepting the post of Surveyor to the line at a fee of £140, including the costs of assistants. He also asked Pease to 'send a suitable person along with me who knows the different Gentlemen's grounds through which we should pass'. In a postscript he added: 'I should like to be paid by the day, which would be 2 guineas together with travelling expences. The other Surveyor that will accompany me I could not offer him much less, I shall also want 2 men to attend to the levelling staves, and 2 to the measuring chain.'

On 27 September Stephenson attended a meeting of the management committee in Darlington, at which his terms were accepted and John Dixon was appointed as the 'suitable person' with local knowledge to accompany him. At long last, events took on an inevitability and a momentum of their own. Stephenson and his team set up their first office in Edward Pease's Northgate house. Favourable weather enabled their survey to be completed much more quickly than he anticipated – it was begun on 14 October 1821 and concluded on the 31st. George Stephenson was ably assisted in the levelling by his son Robert, who was only eighteen years of age at the time and who thus gained his first experience of railway engineering. The outcome was a Report dated 18 January 1822 – 'a masterpiece of work covering every step of Overton's line with Stephenson's own deviations. He innumerated every aspect of the construction of the line.'[5] Stephenson's Report, together with his plans and estimates, were presented to the management committee and accepted by a General Meeting of the shareholders held in the company's office in the High Row, Darlington, four days later, 22 January. His alterations to Overton's route made many gradients less severe, and reduced the overall length of the line authorised by Parliament by nearly 3 miles, and this despite the fact that he brought it much nearer the northern outskirts of Darlington. This brought about a more than proportional saving in the estimated construction costs, as we have seen. On this basis Stephenson was appointed engineer to the railway at a salary of £660 per annum, out of which he was to meet all his expenses and 'provide for the services of assistants', as specified in his letter of 13 June 1821. At the same time he was authorised to begin construction of those sections of the line which conformed to Overton's route pending an application to Parliament for a second Act to take account of the deviations that he recommended. The General Meeting duly authorised the management committee to proceed with that application.

Stephenson made it clear that he could not devote all his time to the S&DR, but the company stipulated that he should spend at least one week in each month on the works. Two resident engineers were appointed under his supervision, John Dixon for the section from Stockton to Heighington Lane, and Thomas Storey, a Northumbrian mining engineer recommended by Stephenson, for the section from Heighington Lane to Witton Park. On Stephenson's advice the line was split for tender purposes into small contract lots of only 1 mile each and the contractors were advised that plans etc. for the sections could be examined at the company's offices in Darlington.[6]

Work began on 13 May 1822 and proceeded apace. It was therefore decided that Inauguration Day, marked by the laying of the first rail at the level-crossing by St John's Well, Stockton on Tees, would be 23 May. The ceremony was performed by Thomas Meynell as Chairman, who was drawn from Yarm in his coach by a party of railway 'navvies', preceded by his own Yarm Town Band. With him were Benjamin Flounders and four local clergymen. On arriving at Stockton they attached themselves to the Stockton procession, led by the Mayor, Thomas Jackson, and the Recorder, Leonard Raisbeck. After the ceremony, which took place to the accompaniment of a peal of church bells, the party proceeded to Stockton Town Hall for the inevitable celebratory dinner. A tablet that commemorates the laying of the first rail has been fixed to the facade of No. 48 Bridge Street, close to the spot where this momentous event took place.

THE ROUTE AND THE BUILDING OF THE LINE

It is convenient to divide the original route of the S&DR as authorised into two stages. The first is the 5.8 mile long section between its beginning at Witton Park Colliery, on the banks of the River Wear, and New Shildon. This we call the hill section because it is traversed by two high watersheds, known as the Etherley and Brusselton ridges, separated by the valley of the River Gaunless, a tributary of the River Wear which it joins at Bishop Auckland. The gradients here are such that locomotive haulage was ruled out from the start. The second, much longer, section, at 19.6 miles, from New Shildon by way of Darlington to Stockton is relatively level, the gradients nowhere exceeding 1:104 – ideal for locomotive haulage. This we call the coastal plain section.

We need to bear in mind that whereas Overton envisaged haulage by horses on the coastal plain section, Stephenson's route was designed from the beginning to accommodate haulage by locomotives. Although both engineers proposed a mixture of horse-haulage and stationary steam-engine-worked inclines for the hill section, even here, in practice, Stephenson substantially deviated from Overton's route and relied to a greater extent on stationary steam-engine-worked inclines in preference to horse haulage, as we shall see.

The first edition Ordnance Survey 6 inches to 1 mile map published in 1861/2 (surveyed 1855–8) indicates that Witton Park Colliery, where Stephenson's line began, was located halfway between Witton Park House (the present East Park Farm) and Witton Park Farm. Rather confusingly, this stretch also served what became known as Witton Park Old Colliery (William Pit) near Witton Park Farm. Old Etherley Colliery, two-fifths of a mile to the east of Witton Park Colliery, was situated, again rather confusingly, in what is now Witton Park *village*. Sadly, all traces of Stephenson's line, and indeed all industrial activity in this area has been obliterated.

From Witton Park Colliery the line ran to Phoenix Row (named for the colliery there) at the foot of Etherley Ridge. Stephenson chose to tackle it directly by means of a pair of inclined planes, ascending and descending, powered by a stationary steam engine at the summit. A clause in the Second S&DR Act of 23 May 1823, which embodied Stephenson's proposals, stated, with reference to the First Act of 19 April 1821: 'whereas it was then considered that one Inclined Plane would be necessary, by means of authorised deviations and alterations the length of the Railway or Tram Road will be shortened by three miles, a greater number of Inclined Planes will be necessary . . .'. This reference to one inclined plane being necessary only in Overton's route is puzzling, since his Plans and Sections plainly include two, one descending to and the other ascending from the flood plain of the River Gaunless.

At the time of the 'S&DR 150' anniversary celebrations in 1975 the Etherley engineman's house and much of the engine-house, together with the associated reservoir, remained *in situ*. Sadly, all this industrial heritage has now been swept away. From the vicinity of Nor Lees House Stephenson's Etherley South incline descended to the boundary of the flood plain of the River Gaunless.

A little further on, Stephenson's line crossed the River Gaunless on the world's first cast-iron railway bridge. This four-span structure was designed by George Stephenson in 1822 and cast at the ironworks of Messrs John and Isaac Burrell in Orchard Street and South Street, Newcastle. Although Stephenson was a partner in the firm at the time, he withdrew on 31 December 1824: it is unclear if this was because of any potential conflict of interest. The original three-span structure was completed in 1823, but after it was damaged by storms and flood on 10 October 1824 a fourth span was added.

Shortly after the Brusselton inclines (*q.v.*) were by-passed in 1856 the bridge became disused. However, when this section of the line from the West Auckland end was reinstated to serve the reopened Brusselton Colliery in 1901, it was replaced by a masonry structure by the NER (as successors to the S&DR). The original bridge was removed and became one of the main exhibits at the L&NER Museum at York a quarter of a century later, after a brief sojourn at Darlington. It now has pride of place displayed outside the National Railway Museum at York. The decking carries a stretch of fish-bellied rail on which sits a contemporary chauldron wagon. The stone abutments of the Gaunless Bridge can still be seen *in situ*, facilitated by a footbridge that spans the river a few yards upstream.

Beyond the crossing of the River Gaunless, Stephenson's routes across the flood-plain ended at 'Bankfoot'. Because Stephenson considered Overton's Brusselton west incline 'dangerously steep', he chose instead to route his Brusselton west incline along the northern slope of the ridge on a much gentler gradient, to a summit level only a few yards to the north of that proposed by Overton. Fortunately, the engine-house and the engineman's house, together with

THE GAUNLESS BRIDGE

The design of the Gaunless Bridge was truly innovative in four respects. First, it combined both the arch and suspension principles of construction. Second, it used both cast-iron – an ideal medium for sections subjected only to compression (e.g. supporting piers) – and wrought-iron, which was more suited to sections subjected to tension and bending (e.g. those carrying the decking). Third, each of the four spans consisted of a pair of wrought-iron fish-belly-shaped girders with cross links of iron cast around them, and extended upwards to carry the timber decking. The girders were cleverly interlocked at each end by a cast-iron boss. Fourth, the bridge was supported by three piers, each comprising a pair of cast-iron columns braced together and splayed outwards. This ingenious construction resulted in the load carried by the bridge being shared, the outward thrust of each arch being counteracted by the inward pull of the suspension element.

The Gaunless Bridge, re-erected outside the National Railway Museum, York.

the associated reservoir, still remain intact to the south of the track-bed here. The engine-house has been converted into a private dwelling. It still bears its S&DR property plate – H 1.[7]

The reservoir at the summit which once served the engine-house has been developed as a nature reserve by Shildon Town Council in association with Durham County Council. It was dedicated on 15 July 1988. In the heyday of the Brusselton inclines a considerable community of workers and their families lived in terraced houses at the summit level. A row of cottages to the east of the engine-house on the north side has long since been demolished, but a row to the west on the south side of the track-bed still remains intact.

From the summit level Stephenson's route descended immediately to the foot of the ridge near the Mason's Arms public house on the Darlington to Bishop Auckland turnpike, by means of the Brusselton east inclined plane. At this point a re-entrant valley fortuitously cut into the otherwise broad and formidable Shildon ridge, between the coastal plain and the valley of the River Wear, of which the Brusselton ridge was an outlier.

The Mason's Arms is situated at the head of a salient of the Durham Plain which signals the end of the hill section of the line. No doubt Stephenson chose to route his line to this point directly from the Brusselton summit level because, with locomotive traction in mind, that secured the longest possible stretch of track in the coastal plain section which followed with its comparatively easier gradients. 'He was at pains to secure an even falling gradient all the way to Stockton.'[8] To further facilitate locomotive traction his route from just beyond the Mason's Arms to the terminus at Stockton consists, for the most part, of six long straight stretches (as numbered below). These were linked with wide curves at each change of direction, so easing the sharp curves with which Overton's route abounded.

(1) From just beyond the Mason's Arms at New Shildon Stephenson's line resumes a south-east course as far as Simpasture where a junction was later formed with the Clarence Railway. By routing his line as far east as Simpasture before turning to run due south, Stephenson circumvented the last spur of the ridge through which Overton was obliged to drive a tunnel. (2) It then passes through Aycliffe Road (later renamed Heighington) and by Whiley Hill Farm, to 'Standalone'. (3) The most difficult stretch for track-laying occurred over Miers (Myers) Flat, a low-lying marshy area between Standalone and Whessoe. Stephenson arranged for hundreds of tons of earth to be dumped into the bog, repeating the process with more earth until a firm foundation was established. (Years later he employed the same technique when constructing the Liverpool and Manchester Railway (the L&MR) over the notorious Chat Moss). Resuming a southerly alignment the line is brought to what was then the outskirts of Darlington just beyond Rise Carr. It then trends first south-east and then (4) east to pass to the north of the then built-up area of the town, but half a mile nearer the centre than Overton's route. The line crosses the Great North Road turnpike on the level, and having bridged the River Skerne it continues for a further 2½ miles on its easterly alignment before once again turning south-east for a short stretch through the hamlet of Fighting Cocks. (5) At what later became Oaktree Junction, Stephenson's route resumes an easterly course.

It was only at this point that Stephenson adopted Overton's line for almost the whole of the remainder of its length to Stockton. Almost certainly, this was because it was here that Overton at long last adopted a straight rather than a meandering route to the terminus. Consequently, Stephenson's line also goes by way of Goosepool to Urlay Nook and then by a slight change of direction to Eaglescliffe (Whitley Springs). Henceforth we are obliged to use the past tense since from this point Stephenson's line diverged from the present line for the first time since leaving Shildon. (6) From Whitley Springs the line turned to run north-east through the grounds of Preston Park, a mansion exactly contemporary with the S&DR as it was built in 1825 for Mr Marshall Fowler. Just short of 2 miles from Whitley Springs Stephenson's line ran north-east for a final 1½ miles to terminate at the S&DR's four wharves on the quayside at Stockton, here called 'Cottage Row'. Just before the terminus the line passed over the Stockton–Thornaby turnpike at St John's Well Crossing. It was here that the first sod was cut by George Stephenson to signify the start of construction on the line on 23 May 1822. (This is despite the fact that approval for Overton's amended route, as surveyed and recommended by Stephenson, was not given until precisely a year later in the Act of 23 May 1823.) There is a fanciful account of this event in M. Heaviside's *History of the First Public Railway*: 'One fine evening when busily engaged with this survey at the

Stockton end, [George Stephenson] said to some of the men by whom he was surrounded "Come give me a spade, let it never be said that we have not made a beginning", and then and there, close to St John's Well, in the Autumn of 1821 [*sic*], he cut the first sod of the new Railway.'[9]

As an aside, it must be said that the reliability of Overton's Plan and Measured Distances has been questioned in the past. Francis Mewburn wrote of his appointment as surveyor: 'This was unfortunate; he was wholly incompetent.'[10] Tomlinson concluded that Edward Pease 'had little confidence in Overton as a civil engineer'.[11] In 1832 Pease commented that 'he was not the man we wanted' – in contrast to Stephenson. Maurice Kirby writes more recently of Overton's 'limitations as an engineer and a surveyor'.[12] The sub-committee that invited Stephenson to undertake a re-survey justified it on the grounds that Overton's survey was insufficiently specific to proceed with the purchase of land or the awarding of construction contracts. It is true that Overton was a better engineer than he was a surveyor. Even with his assistant to undertake the bulk of the surveying, the resulting Plan and Section leave much to be desired – they cannot be taken, by today's standards, as an accurate indication of Overton's intended route. George Stephenson, by contrast, was not only a more accomplished railway engineer, but made use of more professional surveyors to determine and delineate his line. It is not surprising, therefore, that Overton's line as amended by Stephenson and incorporated in the Act of 23 May 1823 not only includes many deviations from that line, but is also presented with much greater clarity.

Since Stephenson was concerned to save expense in forming the cuttings and embankments of his line – 'cuts and batteries' as they were then called – he was critical of Overton's practice. He pointed out that in building the embankments called for on Overton's line, spoil would have to be carted by road a considerable distance, whereas he had arranged his levels so that his embankments could be built from the spoil excavated from neighbouring cuttings and transported in tipping wagons on temporary ways.

Overall, Stephenson's 'main-line' route totalled 24.2 miles compared with Overton's at 26.9 (to be precise 24 miles, 352 yards and 26 miles, 1,540 yards respectively), a saving of some 2.7 miles. Overton's and Stephenson's routes deviated substantially from each other, and for this reason it was necessary to seek Parliamentary approval for the changes.

From New Shildon the line was at first single track throughout, except that passing loops were provided at approximately quarter-mile intervals, initially to allow loaded and empty trains of wagons to pass each other in opposite directions.

Today, when the minutiae of the planning process can delay the construction of major Civil Engineering works, in some cases for years, it may seem surprising that the S&DR was able to complete its line with such despatch. That Parliament was conscious of an overriding need to facilitate the railways' progress, in the light of their potential to advance the nation's economy, is evident from the sweeping powers that were conceded to the railway companies. In the first place, they were empowered to thwart the (sometimes fierce) objections of the landowners and go where they pleased to survey a proposed line. They were then empowered to acquire sufficient land to construct it, once the route was approved, by a form of compulsory purchase that involved the payment of compensation or 'satisfaction'. The First S&DR Act of 1821 set the precedent. The company and its servants were authorised to enter the lands of '*any* Person, Body or Bodies Politic Corporate or Collegiate' to survey and take levels, and then to appropriate such parts 'as they shall think proper and necessary for making, effecting, preserving, improving, completing, maintaining and using the same'. (The strip of land appropriated for building the railway was not to exceed 15 yards in breadth, except at passing loops.) Almost invariably therefore, in frequent and sometimes violent confrontations between landowners' agents alleging trespass and railway servants asserting their right of access, it was the latter who clearly had the law on their side.

Such powers were unprecedented, and coming as they did at the close of an era when the rights of landowners were held sacrosanct, they go some way to explain on the one hand the degree of opposition, and on the other the speed with which the railway network spread throughout the British countryside.

THE WROUGHT-IRON RAIL

As to the physical construction of the S&DR's line, one of the most important questions to be resolved was the type of rails to be employed on the line. As this would be the largest single item of expenditure, apart from the earthworks, it was crucial that the right choice be made. A special sub-committee was accordingly appointed to examine existing railways and tramroads, to obtain tenders and to make recommendations.

In this connection, one of the most important and fortuitous developments, which led to the inauguration of steam-hauled railways at this time, was the invention of wrought-iron. The new material was much lighter, stronger, more malleable and yet less prone to fracture under the hammering of wagon wheels than cast-iron, which had held the stage hitherto. The advantages of wrought-iron over cast-iron in the construction of railways were first suggested by Robert Stevenson and brought to the attention of George Stephenson. Before 1821, however, because they were of simple square section, the use of wrought-iron rails was limited to light, narrow-gauge tramways in mines. Early in 1821 John Birkinshaw, an engineer employed by Michael Longridge (1785–1858) in his Bedlington Ironworks in Northumberland, perfected and patented a method of rolling true wrought-iron rails of 'I' section in lengths up to 20 ft.[13] By applying cams to the operation of the rollers he was able to impart the fish-bellied form that had already been achieved in cast-iron rails first introduced by William Jessop. The rails were supported by chairs and the fish-bellied sections gave greater strength between the chairs, the depth of the web increasing between each pair of chairs from 2 in to 3¼ in the middle. In a report of 28 February 1821 Birkinshaw and Longridge claimed the superiority of a wrought- (or malleable)-iron railway over a cast-iron railway on four counts:

1. Given rails of equal strength, the cost would be less.
2. As the rails could be made in lengths of 9, 12, 15, or 18 ft or longer, the number of joints required would be less, thus reducing the amount of damage caused by the shock of wheels passing over them.
3. The rails could be welded, eliminating the need for joints altogether.
4. Reducing the number of joints and the jolting that resulted would lessen damage caused to engines and rolling stock, through jolting, and the loss of coals from the wagons.[14]

George Stephenson was quick to appreciate the significance of Birkinshaw's invention. On 28 June 1821 he wrote to Robert Stevenson: 'Those [Birkinshaw's] rails are so much liked in this neighbourhood, that I think in a short time they will do away with cast-iron railways. They make a fine line for our engines, as there are so few joints compared with the other.' Not surprisingly, Stephenson strongly recommended Birkinshaw's rails to the S&DR rails sub-committee. Bearing in mind that Robert Stevenson was still advising the S&DR at this time, and that the canal versus railway controversy was not finally settled, Stephenson cannily added: 'I am confident a railway on which my engines can work is far superior to a *canal*. On a long and favourable railway I would stent my engines to travel 60 miles per day with from 40 to 60 tons of goods.'

On 29 December 1821 the sub-committee reported that the lowest tenders they had received were £12 10*s* per ton for wrought-iron, malleable rails as against £6 15*s* for cast-iron. As the weight per yard of wrought-iron rails was less than half that of cast-iron rails of equivalent strength, the cost per mile was approximately the same. On George Stephenson's recommendation the management committee decided that two-thirds of the railway should be laid with wrought-iron rails, and because of financial constraints the remainder, including the passing loops, with cast-iron.[15] The order for the cast-iron rails and all of the chairs was placed with the Neath Abbey Ironworks of South Wales. Michael Longridge's tender for Birkinshaw wrought-iron rails at £15 per ton delivered at Stockton was accepted. They came in 15 ft lengths and weighed 28 lb per yard – they were known to the workmen as ten-stone rails.

On the S&DR each length of rail was supported by four intermediate cast-iron chairs pinned to stone or wood blocks with two 3½ in nails, with a heavier chair at each end enclosing the connection between adjoining rails. Because of financial constraints in the early years, while sandstone blocks were used in the section from Witton Park to Darlington, oak blocks, 2 ft long, 7 in broad and 5 in thick, were used in the remaining section from Darlington to Stockton. These were made from old ships' timbers supplied by Holmes and Pushman, the Portsea, Hampshire ship-breakers, and came by sea to the Tees at the price of only 6*d* each. (Although the cost of the stone blocks was about the same, the cost of transporting them from the Etherley and Brusselton quarries to Darlington and beyond would have been much more expensive.) It is important to note that contrary to some accounts these were not wooden *sleepers* laid crosswise across the track. The original stone blocks can still be seen *in situ* on a stretch of the Brusselton west incline. It was George Stephenson himself who suggested, in his letter to Edward Pease of 2 August 1821, that there were two suitable quarries as sources for stone blocks, 'on the North Side of Brusselton Hill near to the Roman Road close by the ascending plane: this would be a very convenient situation, as the Blocks could be conveyed along the Railway as they are wanted'.

The S&DR was the first railway to use wrought-iron rails to any large extent. When the financial situation eased, almost the whole of that part of the line which was originally laid with cast-iron rails was relaid with wrought-iron rails, although to a heavier specification to cope with the increase in traffic, and the wooden sleeper-blocks were replaced with stone blocks. The few remaining cast-iron rails continued to give trouble for years to come. The S&DR had good reason to be grateful to George Stephenson for his foresight since, as consultant engineer, he had recommended the use of wrought-iron rails for at least two-thirds of the original track. Yet (and this is another instance of Stephenson's integrity) this practice was against his own financial interest since he was a partner and a co-patentee with William Losh, senior partner in the firm of Losh, Wilson and Bell of the Walker Ironworks, Newcastle, which made improved cast-iron rails![16] If the recommendation had been for cast-iron rails, Messrs Losh, Wilson and Bell would undoubtedly have been awarded the contract. Not unnaturally, William Losh was greatly displeased and the incident brought an end to their friendship, and to Stephenson's long-standing association with the firm, dating from 1815. The rift was to have important consequences when it came to ordering the first batch of locomotives for the S&DR, as we shall see.

In 1831–2 the 19½ miles of single-track line from Brusselton Bank Foot to Stockton was doubled, requiring the widening of most of the cuttings. It was probably at this time that the original stone blocks, varying in size up to 21 by 15 in and from 7 to 10 in thick, and weighing about 75 lb – so a man was just capable of lifting one – were replaced with much larger stone blocks 2 ft square and weighing approximately 2 cwt each, to cope with the heavier loads that the track was called on to carry. The new blocks provided a much more stable basis for the track, and larger chairs were used, which were fixed to them by means of *four* fastenings.[17] The later blocks can thus be distinguished easily from the earlier ones provided the number of bolt holes is visible.

It was George Stephenson who was instrumental in securing the adoption of the 4 ft 8 in gauge, which he inherited at Killingworth, for the proposed railway. In his day, there was no standardisation between one colliery and another, so this decision was almost entirely arbitrary. According to some popular historians, the 4 ft 8 in gauge in turn derived from the gauge of Roman chariot and wagon wheels, of which there was evidence in the ruts worn into the roadways of forts along the line of Hadrian's Wall not far distant from the Tyneside collieries. The Killingworth wagonway, the earliest portion of which was laid in 1762, was one of the first, if not the first, in the country, and the 4 ft 8 in gauge was adopted from the beginning. (According to Samuel Smiles, 'the gauge of wheels of the common vehicles of the country – of the carts and waggons employed on the common roads, which were first used on the [colliery] tramroads – was about 4 ft 8 in. And so the first tramroads were first laid down of this gauge.') Ultimately, it was the S&DR's gauge of 4 ft 8 in plus ½ in (1.435 m) that became the standard gauge throughout Britain, and much of the world where the Stephensons, father and son, built the first railways. (There is some dispute as to when and why the extra half-inch was added, but the most plausible explanation is that it was to permit the use of wider flanged wheels by the Great North of England Railway (the GNER).) An Act 'for the regulating the gauge of Railways' was passed in 1847.

The Inclined Planes

Inclined planes were crucial to the efficient operation of the S&DR for the greater part of its independent existence. They were well established on existing railways when George Stephenson contemplated their use on the steepest gradients of the route that he proposed for the S&DR in 1823. He himself had built two self-acting inclines from pit to river many years before, first while employed as brakesman at Willington and then as engineman at Killingworth. In 1822 he incorporated inclined planes on his first commission as a Civil Engineer, the building of the Hetton Colliery Railway. Inclined planes usually occurred in pairs where they were employed to surmount ridges in the path of railways, with gradients on either side too steep for horse traction. A pair consisted of an 'ascending' plane and a corresponding 'descending' plane according to the direction of travel of a train of *loaded* chauldron wagons. It should be borne in mind, however, that they would alternate, *at some stage*, in a cycle of operations, with a train of empty wagons moving in the opposite direction, returning to source. This alternation of trains of loaded and empty wagons was not necessarily on a one-to-one basis, since the lesser weight of empty wagons enabled them to be coupled up in longer formations. (The definition of 'ascending' and 'descending' planes was universally maintained, which led to the anomaly that trains of empty wagons would 'ascend' descending planes, and 'descend' ascending planes.)

Historically there were eight different types of inclined plane in operation, six of which were available to George Stephenson when he came to engineer the line of the S&DR:

1. A simple, single descending inclined plane was worked by gravity, the train of loaded wagons being controlled by brakesmen. Horse traction was used to haul trains of empty wagons back up the incline. Overton's Etherley descending inclined plane was intended to be of this type.

2. A single ascending inclined plane introduces the practice of placing stationary steam engines at the summit, in this instance to haul trains of loaded wagons up the slope. Overton's ascending Brusselton inclined plane was intended to be of this type.

3. *Pairs* of 'self-acting' inclined planes, properly so called, were first introduced in unpowered form. That is to say, the gravitational force, or momentum, exerted by a train of wagons on the descending plane was sufficient to haul a train of loaded wagons up the ascending plane, unaided.[18] Assuming that the descending and ascending trains are of equal weight, the laws of physics, and in particular Newton's second law,[19] dictate that this system can only operate when the descending plane is steeper than the ascending plane. As the descending train meets the level it progressively applies less force to the ascending train. The displacement in height for the descending incline must therefore be greater than the ascending incline – that is, it must descend to a lower level.

4. The next step was to power each one of a pair of inclined planes by means of one or more stationary steam engines located at the summit – not only to haul loaded wagons up the ascending incline but to exert a retarding force to act as a brake on otherwise runaway wagons being lowered down the descending incline. These stationary engines were sometimes called 'winding engines', for the obvious reason that they operated drums on which were wound hempen ropes attached to the front or rear of trains of wagons – the lead rope and the tail rope respectively. On the S&DR these ropes were of unprecedented length – the longest was over 2,185 yards. They called for a novel manufacturing process that involved the taking out of a patent.[20]

5. Although, in the early days, pairs of powered inclines were not self-acting, most of the later ones were. All such systems assumed that the ascending and descending trains were coupled together, either by means of a single rope wound round one drum, or two ropes each wound round a separate drum but mounted on one axle. The single drum worked best with inclines of equal length – a rare circumstance. With inclines of unequal length (whether or not one or two drums were employed) a train on the shorter incline would reach its destination before the corresponding train on the longer incline completed its run. A number of expedients were adopted to enable that train to complete its journey.

In the case of a single drum, its rope needed to be only marginally longer than the longer of the two inclines. When ascending and descending trains reached their destinations, the ropes were detached, the winding gear reversed, and the freed ends (one paid out and the other taken in) returned to their starting point for the cycle to begin again. The procedure was reversed to pass trains of empty wagons in the opposite direction.

6. In some situations, where the topography would have allowed an unpowered self-acting system to operate (Type (3)), although a stationary steam engine was installed it was not used to pass trains of loaded wagons. It was only brought into play to draw

trains of returning empty wagons up one slope of a ridge, in which it had the assistance of the momentum of trains of empty wagons going down the opposite side.

7. It was Timothy Hackworth, the S&DR's first resident engineer, who finally and most effectively addressed the operational problems inherent in pairs of inclines of unequal length. His solution was to mount two drums or 'rope rolls' on one axle, of different diameters proportionate to the lengths of the two inclines. As a result, in this 'Double-Acting' system the rope wound round each of the drums was exactly equal in length to the incline which it served. As a consequence, not only was this ingenious system self-acting but also both trains, on the ascending and descending planes, reached their destination, without interruption, at the same time!

8. The final type of self-acting incline was a stand-alone, one-way installation – effectively a descending incline where the momentum of loaded wagons going down was utilised to draw coupled empty wagons going up the same slope. There were two versions of this type of self-acting incline. The first employed a double track – one for descending loaded wagons and the other for ascending empties. The second version had a midway 'passing loop' after the manner of modern funicular railways. In practice there was a conventional single track with two rails below the passing loop ('meetings') and a hybrid three-rail track above, which accommodated both ascending and descending wagons.[21] This was necessary because whereas only one rope was sufficient to operate the lower section, two ropes each with a row of guiding sheaves were required in the upper section. The third middle rail was laid between the two parallel rows of sheaves and was common to the ascending and descending tracks. A three-rail system was preferred to the four-rail two-track alternative since it saved much valuable space.

In general parlance the term 'self-acting' now tends to refer to this type only, because of its predominance in modern times. Although it did not feature on the Etherley and Brusselton inclines, it was later widely employed throughout the S&DR network.

We can summarise the eight different types of inclined plane that have emerged from this survey as follows:

1. Single Unpowered Descending Incline
2. Single Steam-powered Ascending Incline
3. Pair of Unpowered Self-acting Inclines
4. Pair of Steam-powered Inclines
5. Pair of Steam-powered Self-acting Inclines
6. Pair of Part Steam-powered, Part Gravity-assisted Self-acting Inclines
7. Pair of Double-acting Steam-powered Inclines
8. Stand-alone One-way Self-acting Inclines.

All but the last two were available as options for George Stephenson when he came to engineer the S&DR. His early advocacy of stationary steam engines to power the inclines that he incorporated on his route is indicated in the Second S&DR Act of 23 May 1823. They are specified for the first time in that historic section which also sanctioned the use of 'moveable engines' – locomotives – for the first time: 'It is expedient and necessary for the Proprietors to erect steam engines or other proper machines in certain places at or near the

Railways or Tramways... to make, erect and set up one permanent or fixed Steam-Engine or other proper machines in such convenient situation at or near each of the inclined planes.'

In the event, Stephenson realised Overton's intention and initially installed a single unpowered descending inclined plane on the southern side of the Etherley ridge (Type 1). He chose to power the corresponding single ascending inclined plane on the northern side (Type 2), and the pair of inclines that surmount the Brusselton ridge (Type 4) with stationary steam engines situated at the summit, according to the terms of the Act.

It may seem surprising that Stephenson chose static steam engines to power the first section of his line at a time when he was strongly advocating locomotive power for the rest. However, not only was the topography markedly different but Stephenson himself had a vested interest in the manufacture of static steam engines (as well as locomotives) through his partnership in the firm of Robert Stephenson and Company at their Forth Street Works in Newcastle. At the beginning of September 1823, father and son set out on a long business trip taking in London, Bristol, Shropshire and Ireland which was mainly concerned with the commissioning of stationary steam engines and boilers.[22]

No modern scruples about the ethics of being an 'interested party' inhibited George Stephenson as engineer of the S&DR line from arranging for an order to be placed with his own firm for two, 30 hp engines to power the Brusselton inclines, at a cost of £3,482 15s, and two 15 hp engines to power the Etherley north ascending incline with its lighter load, at a cost of £1,982 15s. Together with the engines, orders were also placed for the associated boilers and winding gear, which initially consisted of drums mounted on *vertical* axles. The design for the Brusselton system was drawn up, and the machinery installed, by George Stephenson's son Robert, before he took up a post in South America in June 1824 – when he was only twenty years of age.

The Brusselton engines were erected first. Unfortunately, the Etherley engine did not reach the site until mid-September 1825, uncomfortably close to the proposed opening day on the 27th. Timothy Hackworth had only been in post for a few months as 'Resident Engineer... particularly to have the superintendence of the permanent and locomotive engines'. In the former capacity he spent many long hours and sleepless nights at Brusselton and Etherley in the days preceding the opening, ironing out last-minute teething troubles. It was largely because of Hackworth's untiring efforts that the stationary steam engines performed without a hitch on that memorable day. Within months, however, he found it necessary to carry out major repairs to the Brusselton double engine, which had proved unequal to the task. Trains of wagons proved difficult and slow to haul. In addition, the irregular coiling of the ropes on the vertical drums caused operational problems that led to breakage and the consequent destruction of wagons. Not surprisingly, the system as a whole was unreliable, and on occasions ground to a halt.

Hackworth understandably felt that the problems brooked no delay in their resolution. In the absence of Robert Stephenson abroad, he wrote to Edward Pease with a suggestion that a horizontal drum should be installed instead of a vertical drum, foreshadowing his invention of the double-acting system. 'I am justified in saying we can, with one horizontal drum, and one rope, "flit over" more wagons than we are doing at present, besides the saving in capital – being able to do with one rope instead of two.' Although not all of Hackworth's suggestions were adopted, as we shall see, he was encouraged to develop the double-acting system (Type 7), which he installed at Brusselton in 1826. The drums or 'rope rolls' were 13 ft 4 in and 6 ft 2 in in diameter, proportionate to the Brusselton west and east inclines respectively. A double-acting system, of course, is by definition a self-acting system. To repeat: the momentum of the loaded wagons descending the east incline assisted the engine(s) in hauling the loaded wagons up the west incline. The result was a marked increase in the efficiency of the operations as a whole – 'three times the traffic with half the power'.

In the meantime, the Etherley winding gear had also been changed from a vertical drum mounted on a vertical shaft to a horizontal drum mounted on a horizontal shaft. This enabled a changeover to a self-acting system of Type 6 – a pair of steam-powered, part gravity-assisted inclines. From October 1827 this was again modified by Timothy Hackworth to produce another double-acting system, Type 7, similar to that in operation at Brusselton, but with a pair of 15 hp engines. We have a description of the arrangements at Etherley in 1829 from the extant notebook of the engineer John Urpeth Rastrick (1780–1850). Together with another expert engineer, James Walker, Rastrick was commissioned by the directors of the proposed L&MR to 'undertake a journey to Darlington, Newcastle and the neighbourhood to ascertain by actual inspection and investigation the comparative merits of Fixed Engines and Locomotives, as a moving power on Railways . . .'. His notes of a visit to the Etherley inclines on 1 January 1829 include an instructive plan of the summit level with its passing loop, pair of 'B & W engines', gearing, horizontal shaft, double-acting drums, flywheel, etc.[23] The operation of this double-acting system is described in some detail, and concludes with the note that 'Mr Hackworth states that 8 loaded waggons descending the south side will draw up 8 loaded waggons up the north side in 9 minutes'. It is clear that each operation involving two trains of loaded wagons alternated with one involving two trains of empty wagons.

The original 60 hp double-engine at Brusselton proved so inadequate to deal with increased loads that by 1831 replacement became imperative. Timothy Hackworth himself designed a more powerful 80 hp engine which was built by R. and W. Hawthorn. It was a massive affair with cylinders of 3 ft diameter and 5 ft stroke. The fly-wheel was 21 ft 10 in diameter and the shaft 22 ft 11 in long with a diameter of 12 in. The four boilers were equally massive – 43 ft long and 5 ft in diameter. The engine-house and attached engineman's house were located on the south side of the track at the summit and the boiler-house and chimney (since demolished) on the north side. The winding shaft with its two unequal drums spanned the space between, high above the wagons that passed beneath them.[24]

The new engine with its associated works had the capacity to haul six sets of twelve wagons in one hour at the remarkable speed of 15 mph. An alarming incident occurred at its first official inspection. 'A wagon containing several Committee members and railway officers, including Messrs. Joseph and Henry Pease, Hackworth and Kitching was being wound down the slope, when the rope broke. All occupants of the wagon jumped clear, except for Mr. Kitching, whose weight, 18 stone, made him less agile than the others. Fortunately, some workmen on the incline saw what was happening and leaped aboard to apply the brakes, much to the relief of all present.'[25] Despite this set-back, it proved to be an excellent and reliable engine. In one day in 1839 the system achieved sixty-seven 'runs', equivalent to the passage over the Brusselton ridge of 2,120 tons of coal. Not, it must be said, without a great deal of hard graft on the part of the engineman, Robert Young, who worked twelve to sixteen hours a day for a wage of 22 *s* a week, and twenty years without a holiday!

It has been suggested that the Brusselton inclines were converted at some stage from single to double track. This was not the case, but the misunderstanding may have arisen from an assumption as to the outcome of another suggestion of Timothy Hackworth's in the letter to Edward Pease from which we quoted above: 'Moreover, should the trade increase, with ease and safety we can attach another shaft by means of a coupling box, which is not practical in the present form, and *thus run a double plane on the west side*, self-acting plane on the east side. We thereby will be able to double work, probably 60 sets per day, 8 waggons to a set.'

Although it is true that Hackworth's meaning in parts of this letter is obscure, he was quite clearly advocating a 'double plane' or double track for the Brusselton west incline. Self-acting planes with double tracks were certainly in existence at this time. In William Strickland's *Reports on Canals, Railways, Roads etc.*, published in 1826, a plan and elevation of inclined planes 'for Mr. Brandling's Railway from Middleton Colliery to Leeds' clearly shows a 'Roadway for the empty wagons' alongside and parallel to a 'Roadway for the full wagons'.[26]

We may be certain, however, that Hackworth's suggestion was not acted upon. The Brusselton inclines were described in a Report by two visiting German engineers, C. von Oeynhausen and H. von Dechen, compiled in 1826 and 1827:

> The inclined planes at Greenfield and Brusselton are generally constructed with as uniform a slope as possible. This cannot be maintained exactly, but the variation is unimportant. It is more important to lessen the slope gradually at the lower end, and to allow it to run into the horizontal gently. It is equally important to lay the railway horizontally on a suitable stretch on the top of the ridge between the two inclined planes. Therefore, where the two steam engines (i.e. stationary engines) stand, the railway is horizontal for a length of about 300 to 360 ft, and it is laid double to facilitate the passage of wagons.
>
> On the inclined planes themselves there is only a single line, as on them there is only one-way traffic.
>
> The lines on the inclined planes are not straight, but make several turns, as the local conditions necessitate. The rope by which the wagons are drawn up or let down runs over guiding sheaves by which any desired turn may be negotiated with ease, provided that it is not too abrupt.
>
> The guiding sheaves stand in the middle of the track about 24 ft apart, and the number of them upon the long inclined planes is therefore considerable.
>
> The shortest of the inclined planes is 825 yd, and the longest 2185 yd. In order to indicate to the stationary-engine attendant at these considerable distances when the train ought to start, a tall signal post is used, at the top of which a disc is turned in an appointed direction. As it is difficult to see these signals at so great a distance, a telescope, continually pointed towards them, is placed beside the seat of the engine attendant, and through this he must look from time to time. When it is not possible to see the signals from the engine house, then long pulling-wires must be employed, but this presents no difficulty.[27]

Once trains of loaded wagons reached the level summit stretch between pairs of inclines, as at Brusselton, how were they moved forward to the descending plane? Although we have no contemporary accounts, it is probable that the S&DR made use of the same system as its successor the NER used elsewhere of having 'kips'. These were humps at the end of the ascending tracks which lifted the wagons above the summit level. As they moved onto the down slope of the hump the rope slackened so that it could be 'unhooked'. The wagons then ran forward under their own momentum to the beginning of the descending plane, to be hooked onto its restraining rope, without stopping.[28] It was Timothy Hackworth who invented the discharge hook or 'dog', a simple and ingenious instrument for detaching or attaching the rope to the wagons as they moved along. He also invented an improved version of the drag-frame or 'cow', which was attached to the last wagon of a train ascending the incline.[29] This was another ingenious device for throwing wagons off the plane if they threatened to overrun the bank-head or run back, should the rope break under the strain, and crash into wagons waiting at the foot of the incline. There is a much-reproduced watercolour of the Brusselton engine-house viewed from the east, as it appeared in 1875.[30]

In passing, the first mention in literature of a railway signalling system is the engineers' account of the signal posts and discs located at the foot of the Brusselton inclines. We may fairly claim, therefore, that railway signals were first employed on the S&DR, even if their primary purpose here was not to alert an *engine-driver* on the move, but a stationary *engineman* at the summit! They were supplemented by an audible signal consisting of a bell or rapper at the top of the incline, connected by a rope or wire to a drum at the incline foot. In addition to the necessity of informing the engineman when trains of wagons were ready to begin their ascent, it was also necessary that he should know whereabouts they were en route. Tomlinson describes an ingenious method of ascertaining the position of the wagons on the Etherley inclines: 'Thomas Greener, the engineman, constructed a small (coupled) model of the inclines, the working of which corresponded exactly with that of the originals.'[31]

In Timothy Hackworth's letter to Edward Pease he proposes 'a self-acting plane on the east side' in conjunction with the double plane on the west side. Although his meaning is not precise, it is possible that he envisaged an incline worked only by gravity (Type 1), although that is unlikely in view of the hazards involved. There are reports of brakesmen on free-wheeling wagons lighting their pipes from sparks from the brakes, which led to a rule which strictly prohibited the practice. It is more likely, however, that this relates to the operation of the Etherley south incline in its early days.

Although, as we remarked at the beginning, the inclined planes were essential to the efficient running of the S&DR main line, they were never a real success – despite improvements brought about by Timothy Hackworth and others. Delays caused by breakdowns and accidents made them expensive to operate. As Robert Stephenson and Joseph Locke pointed out in *Observations on the Comparative Merits of Locomotives and Fixed Engines*[32] (published in 1830), the failure of one link in the chain of stationary haulage engines would bring all the traffic on the line to a standstill.

At one time or another there were thirteen inclines, or pairs of inclines, on the S&DR system with a total length of some 13 miles[33] – powered by gravity, stationary steam engines, or a combination of both. There were pairs of inclines – surmounting a ridge, one ascending, one descending – at Brusselton, Etherley, High Shildon (the Black Boy branch), Mount Pleasant (Stanley) on the S&DR Dearness branch, and Carr House on the Stanhope and Tyne Railway. Also on the S&TR at the Stanhope end there were two ascending steam-powered inclines that immediately succeeded each other at Crawley and Weatherhill. Further east on this line there were two successive descending inclines at Park Head and Meeting Slacks, followed by the single descending Nanny Mayor's incline, and the special case of the pair of inclines that negotiated Howne's Gill. The S&DR complement was rounded off by the Sunniside incline on the Weardale Extension Railway and the Howden incline south of Crook on the Bishop Auckland and Weardale Railway.[34] On associated lines there were two inclines at Rookhope, one on either side of the valley, on the Weardale Iron Company line, and finally one on the private Black Boy (Gurney Pit) Colliery line, which diverged from the main Black Boy branch.

In 1859 the 80 hp Brusselton engine was sold by private contract together with six others 'lately in use on Inclines and upon the S&DR'. From 1856 locomotives reigned supreme throughout the system and the era of the inclines with their stationary steam engines was no more.

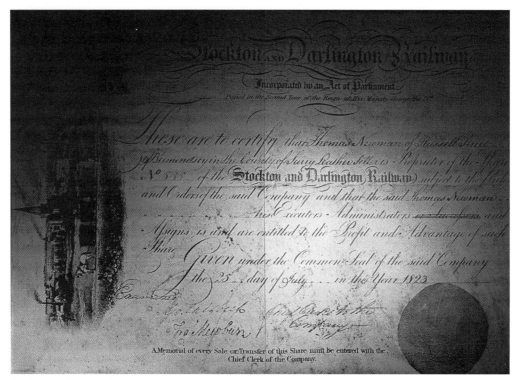

An S&DR share certificate.

LOCOMOTIVE POWER

As the building of the line progressed, attention turned once more to the question of motive power. For four-fifths of the line, east of New Shildon, the choice lay between horse power, locomotives, or a combination of the two.

With the benefit of hindsight it may seem surprising to us that there should have been any question as to the desirability of adopting steam locomotives as the sole source of power. At the time, however, it was by no means generally accepted that steam power would prove superior to horse power. Evidence for this ambivalent attitude is found in the wording of the Enabling Acts, particularly the first, already noted, and the fact that the company's seal depicts a horse-drawn and not a locomotive-drawn train of chauldron wagons. In addition, the first share certificates carried a vignette that also depicted a typical horse-drawn wagonway of the period, which is no doubt what the shareholders at first expected to get. A horse draws three chauldron wagons with an attendant walking behind, against a background of Darlington town centre, seen from the east, in which St Cuthbert's Church, Pease's Mill, Stone Bridge and the Town Hall feature prominently.

It was probably the continuing high cost of fodder for horses initiated by the Napoleonic wars (in contrast with the price of corn, which had fallen dramatically when they came to an end), in addition to its greater efficiency in the bulk transport of goods, which caused the mine owners to consider the possibilities of steam locomotive-hauled transport, using a fuel that they produced themselves. This became comparatively cheaper, since the price of coal also fell as the cost of transporting it by locomotive-hauled transport lessened. In short, fuel continued to cost less than fodder.

In the summer of 1822 Edward Pease again visited Killingworth at George Stephenson's invitation to see his locomotives at work and was, if anything, even more impressed. So also was Pease's Quaker cousin Thomas Richardson who accompanied him, and who formed a very favourable impression of George Stephenson's achievements in locomotive design and operation. This was to prove fortuitous since it was instrumental in securing Richardson's support, and that of his business and family connections, for the Bill before Parliament into which the clause was inserted which sanctioned the use of locomotives as motive power on the new line. That Bill became the Second S&DR Act when it received the Royal Assent on 23 May 1823.

However, the S&DR sub-committee had anticipated the outcome when at a meeting on 9 April it directed Stephenson 'to give estimates for Steam Engines on the Main Line of Railway'. The vital question therefore arose: where should the authorised locomotives be built? The severance of relations between George Stephenson and Messrs Losh, Wilson and Bell, and their substitution by the firm of Michael Longridge in his affairs, carried with it one grave disadvantage. Whereas the former had the capacity to build the stationary and locomotive engines to Stephenson's design, which would be required by the S&DR, the latter did not.

In June 1823, therefore, only two months after the sub-committee's directive, George Stephenson took the momentous decision to register a new enterprise, a specialised engineering firm capable of building locomotives, under the name Robert Stephenson and Company,[35] with premises in Forth Street, Newcastle. (A separate foundry was later built in neighbouring South Street which absorbed the Forth Banks Foundry of Messrs John and Isaac Burrell that adjoined it.) The partners were Robert Stephenson, his father George, Michael Longridge[36] and Edward Pease, who were joined shortly after by Thomas Richardson. The initial capital was £10,000 divided into ten shares, of which Edward Pease took up four and the other partners two each. Characteristically, Pease also loaned Robert Stephenson £500 towards the purchase of his shares.[37]

The Forth Street enterprise was the world's first purpose-built locomotive works. The resolution to purchase the site still survives. It is dated 8 July 1823 and bears the signatures of all four partners. The company bears Robert Stephenson's name because, as managing partner, at a salary of £200 per annum, his was to be the controlling hand at the works, even though he was barely twenty years of age at the time. In this respect, there is a striking parallel with Edward and Joseph Pease, father and son, because Joseph also assumed the mantle of responsibility from his father and, by today's standards, at the same comparatively early age.

Robert Stephenson and Company was officially approached regarding the supply of steam locomotives on 16 July 1824. On 16 September an order was placed for two engines at a cost of £500 each. The consequence was that the railway's first engines, later named, appropriately, *Locomotion No. 1* and *Hope*, were built by Robert Stephenson in Newcastle to a design based on the 'Killingworth engines' built by his father George. An order was also placed with the firm for the two stationary winding engines for the Etherley and Brusselton inclines.

GEORGE STEPHENSON AND THE DEVELOPMENT OF THE STEAM ENGINE

Contrary to popular myth, and a prevailing misconception, George Stephenson did not 'invent' the steam locomotive. If that honour belongs to any one man, it would be the Cornishman Richard Trevithick (1771–1833). This is not the place to review the evolution of the steam engine from Trevithick's prototype of 1804 to Stephenson's *Locomotion No. 1* of 1825. Nevertheless, it is necessary at this point to set George Stephenson's achievement in the context of a peculiarly local and quite remarkable concentration of innovation and inventiveness which occurred in the coalfields of north-east England, centred on Tyneside, during this period. George Stephenson was foremost in a small band of resourceful Tyneside colliery engine-wrights who pioneered the development of the locomotive to a point where it was in general more efficient and economic than horse power for hauling trains of wagons, laden with coals, on the privately owned 'wagonways' that led from the collieries to loading staithes on the River Tyne and the River Wear.

The seed bed for innovation in locomotive design was Wylam Colliery, where George Stephenson was first employed and near which, in High Street House, he was born.[38] Christopher Blackett, the proprietor of Wylam Colliery, had news of the successful performance of Trevithick's first engine at Pen-y-Daren in South Wales on 22 February 1804.

Trevithick's invention was at first called a *tram-wagon*. It was only later called an *engine* because a 'steam engine' was part of it, and it is for this reason that locomotives have been known as 'steam engines' to this day. The fame of Trevithick's Pen-y-Daren locomotive spread and very soon reached the Tyne. As a result, in October 1804, Blackett asked Trevithick to send a plan of his locomotive 'with friction wheels' to his agent, John Whinfield. 'The second of the species' was consequently produced at Whinfield's Pipewell Gate, Gateshead, Foundry, superintended by John Steele, a mechanic who had worked under Trevithick in fitting out his first engine at Pen-y-Daren. It was completed in 1805 and,

---

RICHARD TREVITHICK 1771–1833

If any man can be said to have invented the steam engine, it was the Cornishman Richard Trevithick. Sadly, he was a man much before his time, who died unrecognised and in obscurity. Trevithick's first engine was successfully tested at Pen-y-Daren in South Wales on 22 February 1804. Trevithick's patron, Samuel Homfray, owner of the Pen-y-Daren Ironworks, had taken on a wager of 500 guineas with Anthony Hill, a neighbouring ironmaster, who maintained that 10 tons of iron could not be hauled by steam power over the 9½ miles of cast-iron plate tramway from Pen-y-Daren to Abercynon Wharf.[40] Trevithick's engine convincingly demonstrated that it could, and Homfray won his bet! His third engine was built by John Urpeth Rastrick at Stourbridge in 1808. This celebrated locomotive was put on display for the public in Euston Square, London, in 1809. It travelled on a fenced-off circular track and was christened 'Catch-me-who-can' by Mr Giddy's aunt (hence the expression).[41] Trevithick visited Newcastle several times between 1805 and 1808. On one occasion he sought out and visited George Stephenson in his West Moor cottage. Disheartened at his lack of success in England, Trevithick abandoned his development of the steam engine and travelled to South America where he took up posts as engineer in the silver mines of Peru and Costa Rica. In 1827 a chance encounter took place between the impoverished and forgotten Richard Trevithick and the successful young engineer Robert Stephenson at an inn at the port of Cartegena. Trevithick stared incredulously at Stephenson and exclaimed, 'Is that Bobby? Why I've nursed him [at Killingworth] many a time.'[42] Stephenson gave Trevithick £50 to enable him to pay his passage back to England. He died in poverty in Dartford in 1867.

according to an eyewitness, on 1 May (when George Stephenson was twenty-three years of age) 'it drew three waggons of coals on Wylam waggonway – I saw an engine this day upon a new plan – it is to travail with the waggons'.[39]

Despite its impressive performance, Christopher Blackett refused delivery of Trevithick's 1805 locomotive, and for a reason which is very pertinent to any study of the development of railways in this early period – the inability of the available track to sustain the excessive weight of the locomotive without irreparable damage. Most of the wagonways of the time had traditional wooden rails, many of them strengthened with strips of hard-wearing beech on the surface – the so-called 'double way'. While they could bear the weight of chauldron wagons, they fractured under the punishment they received from locomotives. Their gradual replacement by cast-iron rails after 1807 improved matters, but only to a limited extent. Cast-iron was inherently brittle, and the fact that rails could only be cast in short lengths increased the number of joints that were subject to wear as the engines passed over them, quite apart from the damage from jolting which they sustained in the process. It is to the credit of George Stephenson that he perceived this vital interdependence of locomotives and track and the need to improve the performance of both.

The S&DR was the first to benefit significantly when George Stephenson put his perception into practice. Even in the 1830s locomotive development was dependent upon parallel improvements in the quality of the permanent way, both in terms of the manufacture of the rails themselves and the engineering techniques of construction. It is for reasons such as this that the S&DR, in its early years, was said to be a testing ground (for example, in the use of edge rails) from which the proprietors of other proposed railways would benefit in their turn.

In 1813 William Hedley, the viewer at Christopher Blackett's Wylam Colliery, with the assistance of Jonathan Foster, the chief engine-wright, conducted a series of experiments to overcome the received fallacy of the time that an engine could only move its own weight.[43] The outcome was the manufacture of two Trevithick-type engines, engagingly named *Wylam Dilly* and the eponymous *Puffing Billy*,[44] which were first proved on the Wylam wagonway.

In 1812 Stephenson was appointed engine-wright at the High Pit Colliery, Killingworth, and made his home at West Moor nearby. Shortly after his appointment, Stephenson would be privy to the performance of Hedley's engines on the Wylam wagonway through his friendship with Jonathan Foster. It was at Killingworth, therefore, that Stephenson, following in the path that Hedley had pioneered, was able to develop his own plans for steam locomotives and also to devote time to working out new principles of railway operation and construction. Towards the close of 1813 Sir Thomas Liddell of the Grand Allies ordered Stephenson to supervise the construction of a locomotive for the Killingworth wagonway in the West Moor Colliery workshops. The outcome was the design and manufacture of his first locomotive, *Blucher*,[45] which he called a 'travelling engine', in 1814. On 25 July a great throng of spectators watched it make its first journey, rumbling slowly past Stephenson's cottage at West Moor. The Killingworth railway was laid with cast-iron edge rails so that *Blucher* was fitted with flanged wheels.[46] Her performance on that memorable day, and that of her two successors the *Wellington* and *My Lord*, proved the feasibility of adhesive working on edge rails. 'In its mode of operation *Blucher* ranks as the first "modern" locomotive, reliant on the forces of adhesion via flanged wheels.'

Meanwhile, Hedley's engines were demonstrating serious shortcomings for the work they were called on to perform, largely because they suffered from the prevailing defect of an inadequate supply of steam. It was down to Stephenson's genius that he was able to combine the best features from the engines of Trevithick, Hedley and John Blenkinsop (whose rack-

> ## WILLIAM HEDLEY 1799–1843
>
> A pioneer designer and builder of locomotives, William Hedley was born at Newburn, near Newcastle upon Tyne, and educated at Wylam. He was not yet twenty-one when he was made viewer at Walbottle Colliery, from which he moved to the same post at Christopher Blackett's Wylam Colliery. In 1813 he conducted a series of experiments to refute the accepted fallacy that an engine could only move its own weight. In a letter to Dr Dionysius Lardner in 1836 he wrote: 'I am the person who established the Principle of Locomotion by the friction or adhesion of the wheels upon the rails. And further that it was the engines on the Wylam Railroad that established the character of the locomotive engine in this District, as an efficient and, as put into competition with horses, in the conveyance of coal waggons, an economical prime mover.' The 'engines on the Wylam Railroad' were two Trevithick-type locomotives called *Wylam Dilly* and *Puffing Billy*, which Hedley manufactured. Besides patenting the smooth wheel-and-rail system in 1813, which he laid down on the Wylam wagonway, Hedley also resolved the problem of the excessive weight of locomotives by doubling the number of axles to four. In 1808 be became a ship-owner and in this role, in 1822, he broke a keelmen's strike by placing one of his steam engines on a barge and towing the keels to the coal-staithes 'without the men's assistance'. He died in 1843 and was buried at Newburn on the Tyne.

and-pinion system for the Middleton Railway proved to be a dead end) in the design of a succession of locomotives at Killingworth. His chief improvements on Trevithick's design were: to simplify the transmission of thrust to the driving wheels by coupling the connecting rods directly to crank-pins on the wheels instead of by spur-gears, and by placing the cylinders in the top of the boiler instead of the end; and to minimise side thrusts during each revolution of the wheels by means of cross-beams on the tops of long piston rods. He also collaborated with William Losh, the senior partner in the iron-making firm of Losh, Wilson and Bell of Walker-on-Tyne, Newcastle, to produce the so-called 'steam spring', an arrangement of pistons on each axle bearing the weight of the boiler and its attachments. This cushioned the locomotive against the shocks that were inevitable from a rough and irregular cast-iron permanent way, thus improving adhesion between wheels and rails and cutting down on the number of rail breaks.

Although William Hedley had taken out a patent for certain features of his design on 13 March 1813, it was George Stephenson who, together with Ralph Dodds, the head viewer of Killingworth Colliery, took out the first patent for a 'locomotive' as such, in respect of his second engine, the *Wellington*, on 28 February 1815. On 30 September the following year a patent was registered in the names of George Stephenson and the ironmaster William Losh for (another) locomotive and for 'a method or methods of facilitating the conveyance of carriages and all manner of goods and materials along Railways and Tramways employed for that purpose'. This referred to what is arguably George Stephenson's principal improvement over other locomotives of the period, that is the use of flanged wheels that ran on flat edge rails, which became the standard for *rail*ways, in contrast to plain wheels running within flanged rails, the standard for *tram*ways, which had reigned universally hitherto. The patent also covered Stephenson's invention of the half-lap rail joint that replaced the prevailing butt-joint: 'One rail overlapped its fellow at the joint, and the seating of their chairs was slightly

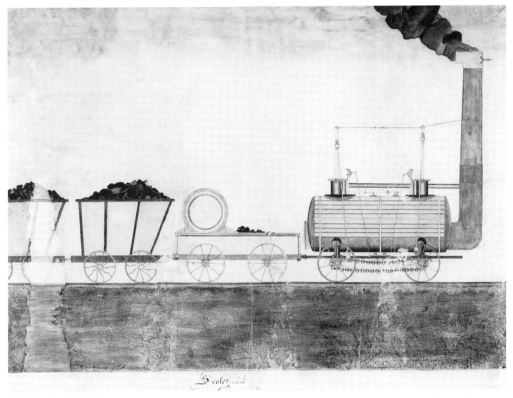

George Stephenson's Killingworth engine, 1816. (National Railway Museum/Science & Society)

curved instead of flat, the rail normally resting on the apex of the curve so that if the sleeper canted this was not transmitted to the rail.'[47] Finally, the patent incorporated Stephenson's invention of the steam spring.

Although, as Maurice Kirby has recently demonstrated, George Stephenson's reputation as 'the father of the locomotive', as portrayed by Samuel Smiles, is misplaced, his real achievement was in advancing the design and construction of the locomotive to its modern form in all its essential details.

Appropriately, it was at Killingworth that a steam engine was first called an 'iron horse'[48] or, alternatively, a 'steam horse'. There is a unique watercolour drawing of a Killingworth engine and train, supposedly by George Stephenson himself, in the Science Museum in London.[49] There is a side and front elevation drawing of 'George Stephenson's patent locomotive engine' in Strickland's *Report on Canals, Railways, Roads etc.* It would be a Killingworth engine of which William James wrote with justifiable hyperbole (in his letter to the shareholders of the S&DR in 1821 after a tour of Tyneside railways): 'The *Locomotive Engine* of Mr. Stephenson is superior beyond all comparison to all the other Engines I have ever seen – Next to the immortal Watt I consider Mr. Stephenson's Merit in the Invention of this Engine.'

It was therefore with an impressive *curriculum vitae* that George Stephenson came to the attention of Edward Pease, and it is now evident why the S&DR in its principal features was largely his creation. George Stephenson's move from Tyneside to Teesside was in some measure the effect and in some measure the cause of the movement of railway development and innovation from the northern part of the Northumberland and Durham coalfield, with its extensive network of colliery wagonways, to its southernmost limit in south-west Durham. In both areas the impetus for development was the need to reduce the cost of coal to the consumer and at the same time to maximise profits for the producer. But while in the northern area those costs and profits were bound up with the 'export trade' from the Tyne and Wear to London, in the case of the southern area and the S&DR, initially at least, they were entirely concerned with the local economy. Our focus of attention now turns, therefore, from the Tyne and Wear to the homeland of the S&DR, and it is appropriate that we should begin with arguably the most celebrated product of Stephenson's genius, and one that literally and physically moved from the one to the other.

LOCOMOTION NO. 1

The design of *Locomotion No. 1*, whose manufacture was begun at the Forth Street Works in September 1824, was a considerable improvement on the 'Killingworth' type, particularly in respect of the drive and the valve gear. Contrary to received opinion from the mid-1920s until recently, there is now evidence that George Stephenson was himself directly responsible for these improvements. Surprisingly, and rather mysteriously, Robert Stephenson resigned as manager of the Forth Street Works at this crucial moment to take up an appointment as resident engineer to the Colombian Mining Association in South America, where he spent

Locomotion No. 1 *on a plinth at Darlington Bank Top Station. (Courtesy of Darlington Borough Council)*

'more than three unprofitable years'. While his motivation remains unclear, it probably resulted from disagreements with his father George Stephenson – and an understandable bid for independence! What is clear is that the Forth Street Works, without a steady hand on the tiller, suffered from mismanagement and fell behind with its contractual obligations to supply locomotives for the S&DR on time. As Edward Pease commented in a typically forthright letter addressed to Robert in April 1826: 'I can assure thee that thy business at Newcastle, as well as thy father's engineering, have suffered very much from thy absence, and unless thou soon return the former will be given up . . . what is done is not done with credit to the house.' A reluctant Michael Longridge was appointed an executive partner in Robert Stephenson and Company, 'with the clear objective of infusing the management with greater vigour and efficiency'.[50]

It is probable that Timothy Hackworth, another Tyneside colliery engineer, joined the staff of the Forth Street Works at this time.[51] It can surely be no coincidence that he was also born at Wylam, where his father John was foreman-blacksmith at the colliery. Timothy succeeded him as foreman-smith in 1807 before moving to Walbottle a few miles away, in 1815, and thence to Newcastle in 1824. Timothy Hackworth's practical engineering skills and inventive talent were to have a profound influence on the fortunes of the S&DR. In the light of his subsequent career, it was indeed providential that Timothy Hackworth superintended the construction of *Locomotion No. 1* during his time at the Forth Street Works.

Somewhat belatedly, on 12 September 1825, George Stephenson wrote to Joseph Pease, Edward's son, from Newcastle:

> I beg to inform you that the Improved Travelling Engine was tried here [Forth Street Works] last night and fully answered my expectations. And if you will be kind enough to desire Pickersgill to send horses to take it away from here on Friday it shall be loaded on Thursday evening. I calculate the weight of the engine between 5 and 6 tons.

Accordingly, on completion, *Locomotion*, together with its tender, was rather ignominiously transported on three horse-drawn wagons from Newcastle, and placed upon the line at Aycliffe Road on 10 September 1825. The original station built at this spot shortly after still exists, doing duty as the 'Platform Restaurant'. On the gable wall of the public house, The Locomotion, to which it is attached, there is a large mural in white plaster which depicts *Locomotion No. 1* and some awestruck bystanders on that memorable day. There is a celebrated story of its first steaming on the line written by the man responsible, Robert Metcalf:

> No. 1 came to heighton [Heighington] lane by road, we had to get her on the way: when we got her on the way we pump water into her: we sent John taylor for a lantern and candle to acliffe: when we done that I thought I would have my pipe: it was a very warm day though it been back end of the year: I took me pipe glass and let me pipe: I thought to myself I would try to put fire to Jimmy ockam [James' oakum][52] it blaaze away well the fire going rapidly: lantern and candle were no use so No 1 fire was put to her on line by the pour [power] of the sun.

(In many histories this story has wrongly been applied to *Locomotion No. 1*'s steaming on the opening day.)

Locomotion's move from Newcastle to Aycliffe was symbolic of the shift of the centre of gravity of railway development from the Tyne and Wear to the hinterland of the S&DR which we have already noted. The majority of the contractors and the elite of the labour force for the construction of the S&DR were recruited by George and Robert Stephenson at Killingworth. According to Thomas Summerside, one of George Stephenson's early workmates, they took from West Moor the best blasters and tunnel-makers, and among those recruited were George Stephenson's own brothers James and John, 'besides many a workmate of his younger days'. As L.T.C. Rolt has pointed out,[53] on a line of such unprecedented length, Stephenson's small, skilled force could do no more than leaven the larger lump of local labour, and sections of the work which they took up had to be sub-contracted to local men, 'often with dire results'. Despite set-backs caused by such inexperienced labour and adverse weather, the building of the line moved steadily forward.

On 13 July 1823 Stephenson was able to report that 21 miles of the line had been completed, that all the rails and chairs had been delivered, and that the railway should be completed by the following midsummer – 1824. However, because of an exceptionally severe winter in 1823/4, the supply of defective wagon wheels, and vexatious litigation on the part of local landowners, that time-schedule slipped by a whole year. In the meantime, the proprietors were becoming increasingly anxious on account of the parlous financial state of the company, largely because of an overrun of the estimated costs of construction of the railway. 'With these unforeseen expenditures the S&DR Company set a precedent which virtually all succeeding railway enterprises were to follow: there was an inexorable tendency for the expenses in construction to exceed those allowed for in the original parliamentary estimates.' The proprietors were understandably concerned that the railway should open as soon as possible, so that revenue might begin to offset the drain on their financial resources. The company's Third Act, which received Royal Assent on 17 May 1824, ostensibly to replace the sanctioned Evenwood Lane branch by one to Haggerleases Lane, contained a clause that permitted the company to increase its borrowing powers by a further £50,000.[54]

Nevertheless, it was not until July 1825 that George Stephenson was able to announce that the main line of the railway and the Darlington Coal Depot branch would be completed within two months. The management committee acted immediately and set 27 September 1825 as the date for the official opening. On 14 September they issued a 'press release', which was published on the 17th, announcing that the grand opening of the railway 'for the general purposes of trade' *would* take place ten days later. The press notice was followed up by a handbill, dated 19 September, which was widely distributed and which contained a full programme of the proceedings. 'The public [was] being offered an entertainment the like of which had never previously been shown to the world, and the public took the fullest advantage of it, as the Committee had intended. . . . The show promised novelty and possible excitement, and all to be had for nothing. So the people made a gala day of it and assembled in their thousands.'

CHAPTER 3
The Grand Opening

PREPARATIONS

The opening of the railway was announced for Tuesday 27 September 1825. A trial run from Shildon to Darlington was undertaken on the evening of the previous day, with *Locomotion* hauling the 'Company's Coach', which only arrived on the day of the trial from Newcastle upon Tyne. *Locomotion* was driven by George Stephenson's brother James with William Gowland as his fireman. The coach conveyed Edward Pease, his sons Edward, Joseph and Henry, together with Thomas Richardson, William Kitching and George Stephenson. This was the first time a steam locomotive hauled a carriage for the conveyance of passengers on a public railway.

THE INAUGURAL RUN

According to one contemporary account the events of the day began 'so early as half-past-five o'clock in the morning [when] a number of waggons, fitted up with seats for the reception of strangers, and others for the workmen, were fully occupied, and they proceeded [drawn by horses] along the Railway *from Darlington*, towards the permanent steam-engine, situated below Brusselton Tower'.

The proceedings proper began, however, at about 8.00 a.m. when the proprietors and their friends gathered, some at West Auckland and some at the 'Permanent Steam Engine' at Brusselton summit. Here they witnessed the arrival of twelve wagons, at the top of the ascending plane, each laden with 2 tons of coals, and one wagon with sacks of flour, 'the whole covered with people' (an augury of things to come), which were intended to form part of the inaugural train. The wagons with coal came from Witton Park via the Etherley inclines and the wagon with flour was coupled up at the crossing of the St Helen's turnpike where the horses began to draw their train over the Gaunless Bridge towards the foot of the Brusselton ascending plane. From thence it was drawn to the summit at Brusselton at a steady 8 miles per hour by the stationary winding engine.

There are four contemporary sources for the composition and order of the inaugural train, which departed from New Shildon on that memorable morning. As advertised in the handbill of 19 September it consisted of:

1. The Company's Locomotive Engine.
2. The Engine's Tender, with Water and Coals.
3. Six Waggons, laden with Coals, Merchandize, etc.
4. The Committee, and other Proprietors, in the Coach belonging to the Company.
5. Six Waggons, with Seats reserved for Strangers.
6. Fourteen Waggons, for the Conveyance of Workmen and others.
7. Six Waggons, laden with Coals.
(Four additional sets of six wagons each, items 8–11 on the handbill – for workmen and others were separately drawn by horses.)

Some of the items in this table call for comment. The 'Coach belonging to the Company', *Experiment*, was fitted up on the principle of the 'long coaches' with longitudinal side-facing seats which accommodated between sixteen and eighteen passengers. Notwithstanding the distinguished nature of the company it conveyed, the coach was unsprung. It is noteworthy that this was the only vehicle that was covered in the event of inclement weather!

Next in order of prestige were the six wagons 'with Seats reserved for Strangers'. These were nothing more than chauldron wagons normally used for transporting coal, cleaned and spruced up, and adapted for the occasion. The press announcement of the opening had stated: 'A superior locomotive, of the most superior construction, will be employed with a train of convenient carriages [*sic*], for the conveyance of the proprietors and strangers. Any gentlemen who may intend to be present on the above occasion will oblige the Company by addressing a note to their office in Darlington as early as possible.' It was for these 'gentlemen' and 'strangers' that the adapted wagons were provided. (Apparently gentle*women* were not envisaged as passengers.)

Last in this descending order were fourteen wagons 'for the Conveyance of Workmen and others'. These were no doubt worthy folk who were not accounted 'gentlemen' but who nevertheless merited a 'ticket to ride'. All the passengers in these wagons were obliged to stand.

We left the proprietors and their friends at the Brusselton summit, where they examined the winding engines with 'expressions of admiration from everyone, so beautiful is their construction, and so completely do they execute their work'. They then joined a great concourse of spectators at the foot of the descending plane near to the Mason's Arms crossing. As an onlooker recorded: 'The day was fine: I never witnessed so great a crowd.' At the Mason's Arms the crowd watched the coupling of the newly arrived coal wagons with the vehicles already assembled there. The former were divided into two sections and marshalled in the 'procession' in the right order so that, together with the wagon of flour, they could be detached as required at two stops on the journey to Stockton. The men appointed to attend to the brakes, who were distinguished by a broad blue sash over their shoulders fastened with a knot under the left arm, took their places between the wagons. (The rest of the company's workmen were distinguished by a blue ribbon in their buttonholes.)

As a result of the many time-consuming preliminaries, both expected and unexpected, the scheduled departure was delayed by one hour so that the 400 ft long inaugural train left New Shildon at 10.00 a.m. It was headed, of course, by *Locomotion*, with George Stephenson acting as driver, two of his brothers as firemen, and Timothy Hackworth as guard and traffic manager. It is a measure of the sheer novelty of this form of transport that the train was still required to be preceded by a man on horseback carrying a flag to warn of its approach, for safety's sake. We are bound to wonder if the Stephensons objected to this precaution.

The *Durham County Advertiser* carried a long account of the events of the opening day[1] which substantially repeats *verbatim* an eyewitness account written by the Chairman of the company, Thomas Meynell. We cannot escape the conclusion that Meynell wrote his account in the first instance for the newspaper's benefit, and that the writer of the article accordingly drew heavily on it for his material. The journalist's own purple prose is employed to good effect as he describes the start of the journey:

> About this time the locomotive engine, or steam-horse, as it was more generally termed, gave 'note of preparation' by some heavy aspirations, which seemed to excite astonishment and alarm among the 'Johnny Raws', who had been led by curiosity to the spot, and who, when a portion of the steam was let off, fled in a fright, accompanied by the old women and children who surrounded them, under the idea, we suppose, that

some horrible explosion was about to take place; they afterwards, however, found courage sufficient to return to their posts, but only to fly again when the safety valve was opened. Every thing being now arranged, the welcome cry of 'all ready' was heard, and the engine and its appendages moved forward in beautiful style.

The scene, on the moving of the procession, sets description at defiance; the welkin rang with loud huzzas, while the happy faces of some, the vacant stare of others, and the alarm depicted on the faces of not a few, gave variety to the picture.

Astonishment, however, was not confined to the human species, for the beasts of the field and the fowls of the air seemed to view with wonder and awe the machine which now moved onward at the rate of ten or twelve miles an hour. . . . A number of gentlemen, mounted on well-trained hunters, were seen in the fields on both sides of the railway, pressing forward over hedges and ditches, as though they were engaged in a fox chace, yet they could not at this time keep up with the procession.

The train soon picked up speed on a moderate gradient, for 1 mile at 1 in 144 and for a further 1½ miles at 1 in 178. The 8-mile journey to Darlington took two hours, but this included three unscheduled stops. The first two were caused by the derailment of the wagon carrying the surveyors and engineers resulting from a wheel shifting on its axle. After the second mishap the defective wagon was shunted into one of the passing loops and abandoned. In the process 'a man who was standing near received a severe though accidental blow on his side' – the world's first recorded public railway accident. The displaced surveyors and engineers being redistributed among the remaining, already overcrowded, vehicles, the train resumed its journey. The third stop of half an hour occurred near Simpasture when it became necessary to deal with a stray piece of oakum lagging that had managed to foul one of the valves of *Locomotion*'s feed pump. The delays totalled 55 minutes so that the actual travelling time to Darlington, which the inaugural train reached at twelve noon, was 65 minutes, an average speed of just less than 8 miles per hour. However, on some stretches *Locomotion No. 1* achieved a speed of 10–12 miles per hour. On the moderately falling gradient of 1 in 134 on the 1¾ mile stretch between Burtree Lane and Darlington, 'in one place, for a short distance', it even achieved an unprecedented 15 miles per hour, 'when it was wished to ascertain at what rate of speed the engine could travel with safety'. Nevertheless, the anticipated schedule was now an hour out of gear.

One incident that has 'passed into history' occurred during the journey from New Shildon to Darlington. Early in the day much of the local population was being conveyed along the railway itself in unauthorised wagons belonging to enterprising local carters. (Many had been employed on the line during the construction period.) Later in the morning a rivalry developed as to which of these wagoners should follow immediately after the main train, and before the company's horse-drawn procession. Considerable initiative was shown by a certain J. Lanchester, who went down the line with his wagon to the first 'passing loop'. From this vantage point he was able to come out onto the single track immediately behind the last of the inaugural train's chauldron wagons after it had passed, and so take up the prized leading position, which he retained all the way to Stockton.

A description of this part of the journey, extracted from the *Newcastle Courant* of 1 October 1925, is particularly vivid:

Nothing could exceed the beauty and grandeur of the scene. Throughout the whole distance, the fields and lanes were covered with elegantly-dressed females, and all descriptions of spectators. The bridges, under which the procession in some places darted with astonishing rapidity, lined with spectators cheering and waving their hats had a grand effect. At Darlington, the whole inhabitants of the town were out to witness the procession.

Taking into account the abandoned wagon, it was a train of thirty-four vehicles which approached Darlington and stopped just beyond the junction with the Darlington Coal Depot branch, approximately where Darlington North Road Station now stands. This allowed the rear six wagons containing coal to be unhitched and drawn by horses to the branch terminus alongside the Great North Road turnpike. There the coals were distributed free to the poor of the parish – a telling example of the proprietors' early appreciation of the importance of what we would now call good public relations. Some near-contemporary illustrations of the inaugural train passing by Darlington depict the coal depot in the background, an elevated structure composed of a number of coal drops.

There is an element of ambiguity in the statement 'The Company's Workmen to leave the Procession at Darlington', in the handbill of 19 September. Does this refer to the workmen who were conveyed in the fourteen wagons of the inaugural train, item 6 in the list on p. 00, or the workmen in the four sets of six horse-drawn wagons, items 8–10, which followed on? Since 'procession' is invariably used of the steam-hauled inaugural train, it would surely be the more privileged workmen who disembarked at Darlington. We do know, however, that the twenty-four horse-drawn wagons followed the six wagons of coal down the Coal Depot branch to its terminus. Both sets of workmen were given tickets that specified at which 'House of Entertainment' they were to dine on 'victuals and ale', at the company's expense. 'The Company take this opportunity', announced the handbill, 'of enjoining on all their Work-People that Attention to *Sobriety* and *Decorum* which they have hitherto had the Pleasure of observing.' One wonders if these paternalistic Quaker sentiments were honoured on the day. We have it from Robert Metcalf's account of his working life on the S&DR that the contractors to the second engineer John Dixon and their labourers 'on the open day dine'd at the Three Tuns, Darlington'.

Some of the ticketed passengers left the train at Darlington and their places were taken by others who were only too eager to travel on the second stage of the journey. These newcomers were in addition to those for whom the company had made provision to travel from Darlington to New Shildon in the early hours of the morning for the start of the journey, as noted above.

At Darlington there seems to have been a change of plan. Originally, it was intended that a group of workmen, 'to whom tickets are especially given', should travel on and disembark at Yarm, that is at a point on the 'main line' near what is now called Allen's West Station. It seems that their places were to be taken by members of the Yarm Brass Band. According to Thomas Meynell's account, however, at Darlington two additional wagons containing 'Mr Meynell's Band' were attached to the rear of the company's coach. ('Mr Meynell's Band' was one and the same as the 'Yarm Brass Band'.) As Chairman of the company, he, and possibly he alone, would have the authority to effect the change of plan, and the means to bring the band from Yarm to Darlington. The intention may have been to avoid a time-consuming stop at Yarm, and to entertain the passengers en route from Darlington, 'playing at intervals cheerful and appropriate airs', above the noise of the train. In the event, there *was* a stop at Yarm (Allen's West) 'of unusual length, where some coal waggons were left'. (The Yarm Coal Depot branch itself was not opened until the following month – 17 October 1925.) All except two old ladies, 'whose infirmities, or prejudice, or both combined, prevented them', turned out to witness the advent of the iron horse here.

Incidentally, there is another ambiguity about the provisions made for these workmen in the handbill – 'for whom Conveyances will be provided, on their Arrival at Stockton'. This probably means that they also were intended to dine at 'Places of Entertainment' at Stockton at the proprietors' expense, but we are left in the dark as to how they were expected to get from Yarm to Stockton – by rail or by road?

During the stopover at Darlington, *Locomotion*'s tender was restocked with fuel and water. (There was one more stop to take on water, at Goosepool.) The inaugural train then began the second stage of its journey to Stockton at 12.38 p.m. Shortly after this, it crossed the River Skerne on one of the most impressive features of the line, a bridge in the classical style, the construction of which was witness to one of the rare occasions when George Stephenson was overruled by the proprietors. After Stephenson had prepared a design for a bridge, on 23 April 1824 they suggested that he should consult the well-known Durham architect Ignatius Bonomi (1787–1870), son of the equally celebrated eighteenth-century artist Joseph Bonomi, with a view to securing a joint plan for building the bridge and securing the foundations. Stephenson was not amused by the proprietors' apparent lack of confidence in his abilities and ignored the request. (This is the only recorded instance of any slight rancour between Stephenson and his employers.) A reminder followed and Stephenson dutifully presented his plans for Bonomi's approval. It was because Bonomi proposed some alterations that speculation has arisen as to whether the design of the Skerne Bridge should be credited to Stephenson or Bonomi. The S&DR's records shed little light on the debate. One of the few, but inconclusive references is contained in the sub-committee minutes of 11 June 1824, where the proposed bridge is the subject of a resolution: '. . . that the secretary do write to Mr Bonomi and request his professional assistance in executing the same'. (By this time Stephenson had abandoned his plans for an iron bridge, partly because of the high price of iron at the time, and partly because of the reluctance of iron founders to tender for work of that kind.) The sub-committee minute probably refers to the fact that when Stephenson was called away to Liverpool in the spring of 1824 Bonomi was instructed to superintend the building of the bridge. In the light of its classical grandeur, we may safely conclude that Bonomi's 'alterations' amounted to a complete redesign. They were approved by the proprietors and on 6 July the foundation stone was laid by Francis Mewburn. It is a measure of the historic importance that the company rightly attached to their undertaking that the S&DR was the first railway in the world to engage an architect.

The design incorporates a distinctive Italianate-style stone arch that spans the river itself, with two narrow land arches on either side, for use by pedestrians, built into extensive curving wing walls. When the doubling of the line was undertaken in 1831/2, it was necessary to strengthen the existing bridge and add a 'new' section on its north side.

From Yarm the train again picked up speed and after just over half a mile closed with the Yarm to Stockton turnpike at Whitley Springs, near to the present Eaglescliffe Station, and 2¾ miles from the terminus. The route here turned north-east to run alongside the road for a further 2 miles. 'Where the railway crosses the Stockton Road', wrote Meynell, 'a train of carriages were waiting the arrival of the procession, and accompanied it to where the railway separates from the turnpike.' The proximity of the road was clearly the reason why such a dense crowd of spectators had assembled here. Along this stretch, the train frequently achieved a speed of 12 miles per hour, possibly in an attempt to make up on lost time.

It was also along this stretch that the celebrated contest between *Locomotion* and a stagecoach took place:

> The procession was not joined by many horses and carriages until it approached within a few miles of Stockton. At one time the passengers by the engine had the pleasure of

accompanying and cheering their brother passengers by the stage coach, which passed alongside, and of observing the striking contrast exhibited by the power of the engine and of horses; the engine with her six hundred passengers and load, and the coach with four horses and only sixteen passengers.[2]

This incident is graphically portrayed in a much reproduced painting by Terence Cuneo that was completed in December 1947.[3] It is alive with movement – of the inaugural train, the straining horses, running youths and barking dogs. History does not record which mode of transport won the epic contest on that day, but with hindsight we can have no doubt which was ultimately the victor.

All accounts of the opening day testify to the numbers of horses which were present along the route, many of the riders attempting to race the train. The reports explode the myth, so assiduously fostered by the landowners who opposed the coming of the railways, that the engines would 'frighten the horses'. At the approach to Stockton: 'Numerous horses, carriages, gigs, carts and other vehicles travelled along with the engine, and her immense train of carriages, in some places within a few yards, *without the horses seeming the least frightened*.'

One has to wonder too, what the emotions were of the outrider who heralded the approach of the train, with his red flag held aloft, when it began to overhaul him at the thunderous rate of up to 15 miles an hour. Did he prudently get out of the way at this stage? Just as his presence on the opening day was a symbol of the superiority of the horse over mechanical transport in the minds of the establishment at the time, so the jettisoning of the short-lived outrider as a necessary adjunct to the steam train was an equally potent symbol of the superiority of rail over road transport in the years to come.

From the point where the turnpike left the railway the gradient fell away for the last mile into Stockton. It is recorded that George Stephenson here gave the engine its head to achieve an unprecedented 15–16 miles per hour. It is doubtful if *Locomotion* could have withstood such a speed for any length of time, the smoke-box reputedly glowed red hot. It was along this stretch that the second accident of the day took place. According to the *Durham County Advertiser* (here based on Thomas Meynell's account):

> The crowd upon the railway in the immediate vicinity of the town was alarmingly great. The road beyond that allotted to the waggons being very narrow, and the engine and its appendages moving on the descent at the rate of fifteen or sixteen miles an hour, the most serious apprehensions were entertained that some accident must happen, for it was found quite impossible to restrain the enthusiasm of the multitude. These fears were unfortunately but too well founded, for a keelman, named John Stevens, who clung for some time to the waggon immediately in front of the coach, at length stumbled and fell, and one of the wheels of the coach passed over one of his feet, which was dreadfully crushed, and it is believed, amputation must be resorted to to save his life.

As the inaugural train approached Stockton:

> A most lively scene presented itself to those who occupied the coach and waggons, the bridge and the neighbouring roads and fields being literally crowded with spectators, who testified their satisfaction with approving shouts, and by the waving of hats and handkerchiefs.

Amid such uproar, *Locomotion* with its attendant train rolled to a halt at the company's wharf at Stockton at 3.45 p.m., at the end of its 20-mile journey from New Shildon. It had covered the 11.8 miles from Darlington in three hours and seven minutes – a disappointing average speed of some 3.8 miles per hour. When account is taken of the stops at Goosepool and Yarm, however, the average speed when travelling was probably nearer the 8 miles per hour of the first part of the journey.[4]

The train and its passengers were greeted by the sound of the band playing 'God Save the King', the ringing of church bells, a 21 gun salute fired from seven 18 pounder guns, and 'three by three stentorian cheers' from the riotous crowd. The proprietors, and such of the nobility and gentry who had honoured them by their company, accompanied by the men who attended to the wagons and preceded by the Yarm Brass Band transported there for the occasion, processed two by two to the Town Hall in the High Street, where a sumptuous banquet, presided over by Thomas Meynell, was served to 102 guests at 5 o'clock, two hours behind schedule. No one seemed to mind the delay! Significantly, among the guests on the top table were the Chairmen of the proposed Liverpool and Birmingham Railway and the L&MR. There were in all twenty-three toasts, each accompanied by appropriate music, among them one to the King, after the playing of the National Anthem, to the Royal Family (to the tune 'Hail Star of Brunswick'), 'Success to the Stockton and Darlington Railway' (appropriately to the tune 'The Railway'), with three times three cheers. Fittingly, the last toast was to the company's surveyor, George Stephenson esq., with three times three cheers, and more. Dinners were dinners in those days! The proceedings did not end until nearly midnight when the celebrants were glad to get out into the fresh air, and to travel home in conveyances that the proprietors had thoughtfully provided.

As a poignant footnote to that historic inaugural day, Edward Pease, whose vision and determination had in large part brought it about, was unable to attend. His nineteen-year-old son Isaac had died that very morning.[5] (Both Isaac and his sister Mary died of consumption.)

There is no doubt that the proprietors regarded the opening day as a great success, which vindicated their careful preparations, despite the unforeseen mishaps and improvisations along the way. 'The Opening of the S&DR was marked by a carefree atmosphere of gaiety, *optimism*, and goodwill.'[6] The inaugural run exceeded the proprietors' best expectations. There is no doubt also that as a consequence they entertained the highest hopes for the railway's commercial potential. Francis Mewburn noted that 'An export trade in coal to London is to be attempted, if it succeeds the Railway speculation must be very profitable.' Thomas Meynell concluded his account:

> The prospects of the Company are now most flattering, whilst they have the satisfaction of seeing the price of coals reduced one-third to the public. They have now the strongest grounds to effect a much larger tonnage to pass on their road than was originally anticipated. An export trade is *now certain*, for an order is already contracted for 100,000 tons of coals annually, for five years, by *one house alone* in London, the produce of which alone will more than pay 4 per cen. on their whole expenditure. The shares are now valued at £40 premium each. Plenty of purchasers, but no sellers.[7]

It was to the absent Edward Pease that the company had cause to be grateful for that.

How Many Passengers?

It is helpful to distinguish between 'official' and 'unofficial' passengers. The former can be divided into three categories:

1. The proprietors and their friends who travelled in the company's coach.
2. The 'gentlemen' who were invited to apply to the company's offices in Darlington, 'and other strangers'.
3. The company's workmen 'and others'.

The company made careful provision for these official passengers by issuing 300 tickets. Each ticket bore the name of the holder and the number of the wagon to which it was allocated – forerunner of our modern system of seat reservation. The company coach seated eighteen passengers. Six wagons were provided for the 'gentlemen and strangers' (seated) and fourteen wagons for the 'workmen and others' (standing room only). Since, together, they would account for the remaining 282 ticketed passengers, that gives an average of fourteen in each wagon, which agrees with their capacity when calculated from the known dimensions of chauldron wagons, and taking into account the fact that there would be slightly more standing passengers in the fourteen unadapted wagons than seated passengers in the six adapted wagons.

'Unofficial' passengers were those opportunists who took matters into their own hands and secured a 'free ride' by sitting on top of the cargoes of coal and merchandise in the goods wagons. It is evident from the accident that occurred as the train approached Stockton that others clung precariously to the sides of the wagons, and the organisers seem to have been powerless to prevent this invasion.

According to the report of the opening day in the *Newcastle Courant*: '[At New Shildon] such was the pressure and crowd, that both loaded and empty carriages were instantly filled with passengers.' One rather disturbing interpretation of this sentence is that the wagons intended for 'official' passengers were at least partly appropriated by 'unofficial' passengers. And how did it come about that some were loaded and some were empty at this stage? The most plausible explanation is that, as we noted earlier, ticketed passengers from Darlington were conveyed thence to New Shildon in advance for the start of the inaugural run. That would be a wise precaution if, as now seems likely, any carriages left empty at New Shildon for occupation at Darlington were liable to be commandeered by a 'riotous mob' at the very beginning of the journey.

If we assume that the 'unofficial' passengers were confined to, and precariously perched on top of, the twelve goods wagons, and that each carried approximately the same number, fourteen, as a passenger 'carriage', that gives an aggregate of 168 unofficial passengers. Together with the estimated 314 passengers in the official carriages we have a total of approximately 482 for this first stage of the journey (compared with 450 given in the *Newcastle Courant* and 553 in the *Durham County Advertiser*).

When the train reached Darlington the 84 unofficial passengers on top of the coal wagons that were detached there would be dislodged. The total is therefore reduced to 398, but augmented again by the 28 or so bandsmen in their two wagons attached at Darlington. On the face of it, that means a reduced load of about 426 passengers onward from Darlington, assuming that some official passengers waiting to embark at Darlington simply took the places of the workmen who were instructed to disembark there.

The situation changed yet again at Yarm. None of these reports mentions a stop here, but without it we cannot account for the slow average speed of about 4 miles per hour between Darlington and Stockton. It may have been the original intention to detach some of the remaining wagons of coals here and run them down the Coal Depot branch to Yarm, as was the case at Darlington, 'to be distributed to the poor of the parish'. This item on the itinerary would have to be omitted from the order of proceedings on the handbill when it became clear that this branch, although sanctioned in the Act, would not be opened in time. Again, there appears to have been a change of plan, and some wagons *were* detached at Yarm, according to a later account. It is probable that they were the six wagons originally intended for the Yarm Coal Depot branch. If so, their cargo of passengers would be dislodged, reducing the numbers to about 332 when the last stage of the journey began.

According to the *Newcastle Courant*, 'several more clung to the carriages on each side'. It is significant that this practice (which it will be recalled led to the accident suffered by the keelman) is only mentioned in the descriptions of the last stage of the journey, and certainly not of the train as it left Darlington. We know that there were eager crowds awaiting the appearance of the train at Yarm, and a stop here, as also when *Locomotion* took on water at Goosepool, would provide an opportunity for the more enterprising, if not foolhardy, among them to attempt a free ride by clinging to the sides of the wagons – cheek by jowl with discomfited bona fide passengers inside. It may even have been possible to do so by running alongside and seizing the opportunity while the train was in motion by the side of the Yarm–Stockton turnpike before the train picked up speed.

The inaugural train was therefore overcrowded when it arrived at Stockton, although not to the extent gratuitously reported in the *Durham County Advertiser*: 'It was ascertained that *nearly 700 persons* were in and upon the waggons and coach attached to the locomotive when it entered Stockton.'

Bearing in mind all the factors involved it is probable that the actual numbers of passengers on the day was no greater than 482 at any stage. The figures given in the newspaper accounts are to a greater or lesser degree exaggerations.

JOHN DOBBIN'S PAINTING: *THE OPENING OF THE STOCKTON AND DARLINGTON RAILWAY*

The history of this celebrated painting is significant. It is signed and dated John Dobbin 1871– that is, it was completed forty-six years after the event it portrays. It is also inscribed 'Opening of the Stockton and Darlington Railway AD 1825 From a Sketch by the Artist'. It is said to have been commissioned by Henry Pease (1807–81), the last surviving son of Edward Pease, the 'Father of the Railways', for the fiftieth anniversary of the opening. We know that John Dobbin was a guest at the jubilee banquet held on 27 September 1875. In 1936 the painting came into the possession of Darlington Borough Council.

JOHN DOBBIN 1815–88

John Dobbin was a native of Darlington, born in a house in Weaver's Yard – by coincidence situated close by Edward Pease's home in Northgate. The site is now occupied by John Dobbin Road, which leads into Weaver's Way – memorials respectively to the man and his birthplace. At first apprenticed to George Spencer, a cabinet maker, Dobbin became a more than ordinarily proficient provincial artist, and although he is remembered chiefly for one painting – 'The Opening of the Stockton and Darlington Railway' – his other works, especially those set in his home town, have great merit.[8] One of his last creations was a mosaic representation of the Last Supper. When it was rejected by the authorities of Westminster Abbey, for which it was intended, he presented the mosaic to his home town. It now stands behind the high altar of St Cuthbert's parish church.

Sketch of the opening of the S&DR, 'Drawn by John Dobbin on the 27th September 1825'. (National Railway Museum/Science & Society)

The painting poses a dilemma when we come to consider it as evidence for the nature of the events on the opening day. Clearly, it would be unreliable if it was simply based on the recollections of the artist, who was only ten years old at the time, and painting forty-six years later. Fortunately, there are two other 'versions' of this painting which can help us to resolve the dilemma. The first is a rather naive, amateur sketch, now in the Science Museum Collection, which is yet superior to what we would expect of a mere child of ten, however artistically precocious he may have been at the time.[9] The writer of an anonymous typewritten text in the Darlington Library Collection concludes that this was the original drawing by John Dobbin, but that 'at a later date, with a further knowledge of drawing and painting', he touched up the sketch and improved much (but not all) of the picture. This is a convincing interpretation, especially since some of the detail, for example the primitive rendering of the locomotive, is just such an 'unimproved' image as a ten-year-old might record. The anonymous writer also claims that the sketch was inscribed 'Drawn by John Dobbin on the 27th September 1825. John Dobbin born 1815, died 1888'. (Alan Suddes, one-time Curator of the Darlington Museums and Art Gallery, is of the opinion that this inscription is dubious and must have been added later, as the second line must have been written later than 1888.)

The other version of this painting is about the same size as the finished picture, that is 63 by 41½ in, but it is executed in a sepia monochrome. It belongs to the National Railway

Museum and is currently on loan to the Darlington Railway Centre and Museum. It is clearly marked 'Enlarged Sketch of the Opening of the Stockton and Darlington Railway A.D. 1825. From a Sketch by the Artist at the Time'. Alan Suddes has pointed out that the words 'Enlarged Sketch of the . . .' and 'at the Time' have been added in a darker ink at a later date: 'Remove these words and we are left with the exact wording on the finished full-colour painting.' He concludes that the sepia version is Dobbin's preparatory working sketch in which he planned the arrangement and composition of the final picture dated the same year, 1871. John Dobbin was by then a mature professional painter, and quite capable of researching other sources for his painting, in addition to his memory, such as the company's handbill of 19 September 1825. However, it is significant that *Locomotion* and her tender are depicted in the painting as they were *after* they were restored in 1857. It is also possible that John Dobbin may have drawn on the work of other amateur artists, who were also present on the day, as well as his own memories.

We are now better able to evaluate John Dobbin's vast and arresting watercolour painting as evidence. It depicts a rural scene with *Locomotion* and her attendant train of wagons, with their waving, festive passengers, in the middle distance, viewed from the south. The artist manages to include not only three of the four flags that were waved from various points along the length of the train, but the actual wording on the third: the company's motto *Periculum Privatum Utilitas Publica*. We know that the motto was repeated on one of the other flags.

In the foreground, where he is able to treat his subjects in greater detail, Dobbin depicts an orderly host of spectators gathered by the banks of the River Skerne, drawn from all walks of life. The gentry are there, some standing, others remaining in their carriages – six are painted into the scene – the better to view the spectacle. Farmers and their families are there, 'up from the country'. Poor labourers and navvies are there, one shouldering his spade. Artisans and their families are also in evidence. Excited children are there, one of whom may have been John Dobbin himself. In the left foreground a little boy holds a toy boat attached to a string, talking to a lady. This is the one major detail missing from the sepia preparatory version. It strongly supports the supposition that John Dobbin, as an afterthought, determined to paint himself in as present on this momentous day, for posterity, and at the same time to pay tribute to his mother.

The painting, overall, is carefully composed. For example, the spectators in the foreground are contrasted with the inaugural train in the distance. One element in particular is highly symbolic. The largest of the spectators' vehicles is not one of the elegant carriages, but a slow 'stage-wagon' prominent in the foreground. The stage-wagon was the goods equivalent of the stagecoach for passengers.[10] It required four pairs of heavy horses to haul it over the inadequate, muddy roads of the pre-railway age, and was equipped with four wide wheels to spread the load. (In contrast, the trainload of nigh on 500 passengers in the background is hauled by a single locomotive.) The front section of a stage-wagon served as the living quarters for the driver and his family, like a gypsy caravan. John Dobbin manages to include all these details, with the driver's wife and daughter seated aloft among the wondering spectators.

Dobbin's choice of vehicles, then, is deliberate and significant. In time, all were to be in large measure superseded by the railways. Although this transport revolution would be apparent to the painter in 1871, at the time of the event that he depicts the spectators would have regarded such a prophecy as preposterous. The contrast, between the road transport of the day and the revolutionary rail transport of the future, is reinforced by a country signpost, near the left-hand margin, with three arms. One points the way ahead to Durham, another back to Darlington town, and the third to the hamlet of Whessoe, which was to lend its name,

Pencil drawing, 'The Opening of the Stockton & Darlington Railway, Sept. 27th 1825', by J.R. Brown. (Science Museum/Science & Society)

not many years after (and certainly by the time the painting was completed) to one of the largest industries the railway was to foster.

One's eye is drawn to Bonomi's classical bridge spanning the River Skerne, in the middle distance, over which the first part of the inaugural train is passing. Regrettably, although it is a historic structure listed Grade II*, it now stands partly obscured by gas pipework and other alien industrial accretions. This railway bridge, the first to be designed by an architect, one of the few that can universally be brought to mind and the first to be made famous by an artist whose own fame, paradoxically, largely rests on this one painting, in which the bridge appears, deserves a better fate.

The Science Museum holds a probably contemporary drawing captioned 'The Opening of the Stockton and Darlington Railway, Sept. 27th 1825', by J.R. Brown, which it acquired from the collection of John Phillimore.[11] The accompanying catalogue entry is instructive:

> Highly finished pencil drawing, 9.75 × 6.5 ins. This unique drawing is in fine condition. The Victoria and Albert Museum considers this drawing contemporary and probably original, that is, not copied from another one. In any case, it is of considerable importance as it shows that a third illustration of the opening day existed. Hitherto there had been only two contemporary representations of the opening of the S&DR, namely the Dobbin sketch and the three-view lithograph (see below). This pencil sketch accords in detail with one of the views of the lithograph but is drawn from an entirely different angle (the north-east), shows clearer and more accurate detail, together with a variation in the crowds and grouping. It shows the train crossing the Skerne, near Darlington, on

Drawing of the inaugural train crossing Skerne Bridge. (From a lithograph in Adamson's Sketches of our information as to rail-roads, *1826. Science Museum/Science & Society)*

its way to Stockton with the 'Band of Music' – supplied by Mr Thomas Meynell, the Chairman – just behind the directors' coach. George Stephenson is driving *Locomotion*.

It also depicts the warning outrider on horseback, holding his red flag aloft, while in the background are shown St Cuthbert's parish church, and the coal drops and coal depot building at the end of the Darlington branch.

The Stephenson Locomotive Society owns a small watercolour painting, 8 by 5 in, undated, artist unknown, and rather primitive in execution, from a viewpoint to the north of, and almost at right angles to, the Skerne Bridge. Although the subject matter is broadly the same as the sketch by J.R. Brown, there are sufficient differences in detail and style to indicate that they were produced independently of one another. Altogether the unknown artist has managed to incorporate on his small canvas a surprising number of detailed features that were part of the scene on that memorable day, which supports the view that this is also a contemporary record. The painting is on loan to the Darlington Railway Museum at North Road Station.

The earliest dated illustration of the opening of the S&DR, however, is the three-view lithograph mentioned above, measuring 20 by 10 in, from the Revd James Adamson's, of Cupar in Fife, *Sketches of our information as to rail-roads. Also, an account of the Stockton and Darlington railway, with observations on railways, etc.*, which was published in 1826.[12] Once again the principal view depicts the inaugural train crossing Skerne Bridge. As a probably contemporary illustration of the opening day, and certainly the earliest that is dated, this lithograph is arguably the most reliable as evidence.

The Skerne Bridge, from an almost identical viewpoint, features in an engraving by J.M. Sparkes which serves as the letterhead of the 'Stockton and Darlington Railway Coaches: Winter (Timetable) of 1837–38'. A locomotive-hauled passenger train of two coaches and an open wagon is shown on the bridge proceeding westwards – a great improvement on the accommodation available in 1825![13]

Promise and Fulfilment

How far did the S&DR fulfil the promise of the opening day, during its early years? It is certain that this enterprise gave the first clear demonstration of the commercial transformation that could be produced by the construction of a railway. The impact of the S&DR on the collieries of south Durham was to be startling.

In the early nineteenth century, land transport facilities were such that a ton of coal that fetched 4s at a colliery near Bishop Auckland cost 8s when carted to Darlington and 12s at Stockton. When, in 1825, Joseph Pease was calculating the probable impact of the S&DR, he estimated that this more economical carriage would reduce the price of coal at Darlington by nearly half and at Stockton by more than half. In the event, the price of coal at Stockton fell from 18s a ton before the advent of the S&DR to 12s per ton afterwards. A little later it fell to 8s 6d per ton. In this respect the success of the project vindicated Edward Pease's predictions, but in fact the value of the line came from a more extended advantage: the coming of the railway provided for the first time a viable access to the lucrative coastal markets for the produce of the mines of inland south Durham. At the beginning of the century, it cost less to move a ton of coal from Newcastle to London by sea than to move it overland from Bishop Auckland to Stockton, a distance of 19 miles. Rail carriage soon produced a reduction in coal transport costs in that area from the road figure of some 4d to 5d per ton per mile to the average cost on Durham coal railways by 1840 of about 1.3d per ton per mile. This startling reduction made a crucial difference to the competitive situation of many inland collieries. This increased profitability brought about by the coming of railways stimulated the sinking of many new collieries.[14]

The S&DR prospectus of November 1818 made no mention of the transport of coal for export. The export potential was only appreciated at the inauguration in 1825 and it was not until January 1826 that the first staithes were ready at Stockton. The first export of coal took place on 26 January 1826, when the sailing ship *Adamant*, with a cargo of 168 tons from Old Etherley Colliery, was towed out of the river by the steam tug *Albion*.[15] Because of the proprietors' lack of foresight when the First Act was drafted, only a halfpenny per ton per mile could be charged.[16] Lack of careful planning was clearly evident in that inadequate space was provided to facilitate the unloading and standing of wagons, and the line was often blocked with fully laden wagons waiting for the ships to arrive. The original wagons could only be unloaded from their ends and not from the bottoms directly into the holds. Nevertheless, in the year ending 30 June 1827, 18,588 tons of coal were shipped from Stockton, and as the number of staithes, wagons and motive power units in the form of either horses or engines increased, so the export capacity improved. Unfortunately, this export potential was retarded not only by the lack of adequate staithes, but also by the limitations of the River Tees for accepting vessels of an adequate size. Vessels of 100 tons or more could not leave the staithes fully laden.

The opening of the S&DR, and its anticipated effect on local coal prices, marked the beginning of a process. The impact on prices of transporting coal by rail, however, was comparatively limited until in 1830 the line was extended from Stockton's unsatisfactory shipping facilities to a deeper riverside terminal that was to be the first impetus of the boom town of Middlesbrough. The S&DR carried 216 tons of coal in the operating year 1828–9, 152,262 tons in 1830–1, and 336,000 tons in 1832–3. The Quaker founding fathers of Darlington had good cause to be satisfied with their investment, and the population in general to be thankful for their foresight.

The success of the inaugural run, together with the practical day by day proving of the system which followed, had another effect of far-reaching significance. It encouraged the hesitant promoters of other schemes, and in particular the L&MR, to press ahead, whereas failure would have put back the onset of the railway era, perhaps by decades. In that respect, the S&DR was in truth 'First in the World'. In the words of Henry Booth, Treasurer of the L&MR, for five years after it was opened for traffic on 27 September 1825 the S&DR was 'the great theatre of practical operations on the railways'.[17] During that time it became a testing ground for practices that, once validated, were adopted by other railway companies which sprang up in its wake.[18] The enterprise was so novel that one gains the impression that the S&DR was 'making up the rules' for railway operation as it went along – in some respects, almost by a process of trial and error. Nevertheless, the company *was* proving that a public steam-hauled railway would literally 'pay dividends'.

Not surprisingly, therefore, from its beginning the S&DR attracted a number of distinguished visitors from other concerns. Present on the inaugural day was William Strickland, who had been sent over as an observer for the Pennsylvania Society for the Promotion of Internal Development – thus sowing the seed for the development of the vast network of railways which sprang up in the USA. Also present were members of the board of the proposed L&MR, which was to send a number of fact-finding deputations in the years that followed. Late in 1825 the great French engineer Marc Seguin, accompanied by his brother Paul and a Mgr de Montgolfier, visited the S&DR at Darlington and Stockton in the course of a journey to this country to survey progress in railways in which he had a commercial interest.

One corollary to its early fame, of course, is that the opening of the S&DR set the seal on George and Robert Stephenson's reputation as railway engineers – the two men, father and son, 'who were together responsible for introducing the locomotive and the railway as a practical means of long-distance transport for passengers and goods'.[19]

CHAPTER 4

The Early Years: Private Risk for Public Service

Common Ground?: Canals, Roads and Railways

The two most significant advances in transport provision prior to the coming of the railways were the development of a system of canals and the spread of a network of turnpike roads.

The labourers who built canals in Great Britain during the eighteenth century were called navvies – short for 'navigators' – because the canals themselves were called 'navigations', in the same sense that rivers were navigations. It may *seem* inappropriate that the labourers who built railways in Britain in the nineteenth century were also called navvies from the beginning. However, in reality, there was a period of overlap between the end of the 'Canal Age' and the beginning of the 'Railway Age', when the same labourers building a canal one day found themselves recruited to build a railway the next. They simply took their familiar title with them.

The word navvy, therefore, nicely symbolises a real continuity or common ground between canals and railways as modes of transport. This continuity was expressed in a variety of ways. For example, George Stephenson, like other engineers of the period, followed the canal precedent when planning his line of railway. For the canal engineers' flights of locks he would substitute an inclined plane, and for the level 'pound' of canal between locks he would have a length of railway that was either as nearly level as possible or falling with the load if, as on the S&DR, the load was all one way. 'On such lengths of railway he championed locomotive haulage.'[1]

Although the canal companies constructed and maintained individual 'navigations', they were 'public ways' in the sense that anyone could make use of them to transport goods or passengers on payment of a toll to the proprietors. By the end of the eighteenth century, this *modus operandi* had spread to the coal wagonways of the north-east, where the practice had begun to develop of throwing open some of the wagonways (notably those of the Grand Allies) to common use on the part of any interested colliery owner. It was this practice of 'common use' which the S&DR inherited and adopted from the beginning. 'At the time of the Opening in 1825 the working of the line bore striking resemblance to established canal practise, insofar as it practically allowed public rights of way.'[2]

On payment of a toll, the colliery owners were permitted to lead their own coals, a custom that was soon extended to the carriage of general merchandise. To contemporary eyes, the most overt example of the company's commitment to canal procedures was its 'laissez faire' policy on passenger traffic.[3] The extent to which the railway proprietors were indebted to the canal proprietors is well illustrated by the 'antediluvian clause' covering restrictions that were placed on the number of operating hours on a daily basis, together with the schedule of tolls, in the First S&DR Act in 1821. The company also followed a well-established practice in the construction of the canal network, when it contracted out the building of the line. Subcontracting in general led to 'an entire sub-stratum of smaller undertakings whose managerial procedures continued to be determined by the precedent of the stage-coach and canal'.[4] Finally, the abortive canal schemes of the eighteenth century were to fulfil an important and useful role after 1800 in informing railway proprietors of likely routes and the attendant engineering problems, as well as the probable sources of revenue.[5]

In the light of subsequent developments, it is one of the ironies of history that the parties who were concerned to initiate an effective transport system between the south-west Durham coalfield and the ports should ever have seriously considered a canal as an alternative to a railway. However, this is understandable, since at that time canals were in the ascendancy. Nevertheless, within a few decades of the advent of the railway companies, they in turn were able to assert their dominance by purchasing many of the canal companies that competed for custom within their territories. They proceeded to eliminate any remaining competition by closing many of their newly acquired canals, to the detriment of the system as a whole. Although this was not directly the case with the S&DR, there is no doubt that its arrival on the scene effectively pre-empted the spread of canals within its boundaries.

Apart from the stimulus of the appalling state of the country's roads, turnpikes were made possible by a fortuitous combination of two novel practices: firstly, a legal obligation that was laid on local authorities to maintain the roads which passed through their jurisdiction, and secondly, with the establishment of the 'Turnpike Trustees' in 1767, the right to levy tolls by private turnpike trusts, at the point of entry, on all vehicles, passengers, pedestrians, driven cattle, and domestic animals which made use of them. In return, the Turnpike Trustees were under obligation to improve and maintain the roads that they took over, and to build new ones. By 1820, of the 125,000 miles of road in Britain, 20,000 were turnpike roads. Turnpike Trusts, therefore, no less than canals, felt threatened by the advance of the railways. An instructive *Notice*, dated as early as 24 February 1819, was addressed to 'The Creditors and Mortgagees of the Tolls arising from the Turnpike Road leading from Darlington to West Auckland . . . re. the proposed Railway from Stockton to the collieries, and its possible injurious effects on their interests'. They were advised 'to apply to Mr. Raisbeck at Stockton or Mr. Mewburn at Darlington (the solicitors to the said proposed Railway) who are authorised to purchase their securities at the Price originally given for the same'.

In the event, the Turnpike Trusts were not so gravely disadvantaged as they feared. Francis Mewburn noted in his diary: 'When the Darlington Railway was projected, the commissioners of the turnpike roads in the neighbourhood opposed this Bill in Parliament, because they thought their roads would be injured by the railway. But mark the consequence. The funds of all the roads increased – so much so that they were able to discharge their debts.' Nevertheless, the rapid development of the railways in Britain brought road building to a virtual halt.

No doubt the railway companies were aware of the power they wielded and the threat they posed. It may seem the more surprising, therefore, that in these early years the S&DR, and others, should give precedence to road as opposed to rail traffic whenever the two modes met and crossed each other. The 'Rules of Engagement' are spelled out in a contemporary poster headed 'Cautions to Engine-men, Waggon & Coach Drivers, etc., on the Stockton and Darlington Railway'.

THE COMPANY'S COACH *EXPERIMENT*

It is evident that from the beginning the proprietors of the S&DR regarded their enterprise as primarily for the carriage of goods, principally coal. Passengers were very much a secondary

The first railway timetable.

consideration and were included in the amended Act of Parliament only as an afterthought. The company must have been surprised at the strength of demand for a reliable passenger service which followed the success of the inaugural run on 27 September. It quickly seized the initiative and announced that it would operate 'The Company's Coach *Experiment*' a mere thirteen days later, from 10 October 1825. The coach would make a daily return journey on Monday, Wednesday, Thursday and Friday; one way Stockton to Darlington on Tuesday, and one way Darlington to Stockton on Saturday. There were to be no Sunday services. The announcement came in the form of a poster, which thus became the world's first passenger railway timetable. (There could be no clearer indication of the railways' ability to generate traffic than the frequency of services in this first timetable. Hitherto 'between Darlington and Stockton there was scarcely passenger traffic enough to effect a reasonably profitable return to the owners of the only (stage) coach which ran three or four times a week on the regular turnpike road'.)[7]

There are two different versions of what the coach *Experiment* looked like on the opening day, and when it inaugurated the first public fare-paying railway passenger service shortly after. On the one hand, in an engraving first published in the 1860 reprint of Samuel Smiles' *The Story of the Life of George Stephenson*[8] (and uncritically reproduced many times since), it is represented as an unsightly, slab-sided 'box' with three small windows in each side and a door at the back where stands a guard – 'a somewhat uncouth machine, more resembling the caravans still to be seen at country fairs' in Smiles' memorable phrase.[9] On the other hand, it is represented as a standard stagecoach set on flanged wheels for running on a railway track, as in Theodore West's *An Outline of the Growth of the Locomotive Engine*, published in 1885.[10] Which version is correct, and how did the confusion arise?

> ## SAMUEL SMILES 1812–1904
>
> The originator of 'Self-Help', prolific author and enthusiastic social reformer, Samuel Smiles was born in Haddington, Midlothian. He was first apprenticed to a firm of medical practitioners and in due course became a GP. Dissatisfied with that calling, he embarked on a lengthy tour abroad. On his return he became Editor of the radical *Leeds Times* in 1838, which signalled the beginning of his career as social reformer. In June 1840 he attended the opening of the North Midland Railway where he met George Stephenson for the first time. In 1845 he was appointed Assistant Secretary to what became the Leeds Northern Railway (the LNR), thus beginning a professional career in the railways which lasted for twenty-one years, including twelve years as Secretary of the South-Eastern Railway from 1854. Smiles was drawn to the writing of biography by a deep-rooted belief that 'concrete examples of men who have achieved great results by their own efforts best indicate the true direction and goals of social and industrial progress'. His first biography was a *Life of George Stephenson* (who had died in 1848) published in 1857, which instantly became a best-seller. This was followed in 1861 by *Lives of the Engineers*, including Boulton and Watt, in three volumes. In volume three, Smiles took the opportunity to add a life of Robert Stephenson to a revised account of the life of his father, George. His fame as the advocate of Self-Help sprang from a lecture, 'Self-Help, with Illustrations of Character and Conduct'. This was enlarged into a substantial published treatise in 1859 which again was a phenomenal success, being translated into many foreign languages.

We know that the original *Experiment* proved unpopular with the public, and this is not surprising if it was as portrayed in the *Life of George Stephenson*. However, an unsprung, 'experimental', modified stagecoach is just as likely to have proved unpopular with the public. Moreover, the box-like vehicle depicted in Samuel Smiles' biography carries no name, whereas, according to a contemporary source, the original *Experiment* had its name, together with the company's motto, painted on the sides.

Whatever the case, the company replaced the original *Experiment* with an improved version bearing the same name in 1827. According to contemporary engravings, the improved version *did* resemble a typical stagecoach of the day. To add to the confusion, one of these engravings was also reproduced in the *Illustrated London News* purporting to be the original. D.H. Watson, who was chief clerk in Timothy Hackworth's office in New Shildon and as a boy was present at the opening of the railway, is quoted as saying that the original *Experiment* was built 'after the pattern of the old mail and passenger coaches, with inside and outside accommodation, those on the outside running great risk of coming in collision with the arches of the bridges'. However, the quotation in fact refers to 'the first coaches' in general, and not to *Experiment* specifically.

It is probable that the original *Experiment* resembled neither the slab-sided box of the *Life of George Stephenson* nor the stagecoach of the *Outline History of the Locomotive Engine*, and it is Theodore West himself who provides the clue. The former, he writes, was a type of conveyance used for a time as a *second-class* carriage.

> This was called a 'tub' by the men: which pretty well explains it, a singularly plain and uncomfortable affair, without springs, and resembling a caravan. The first covered

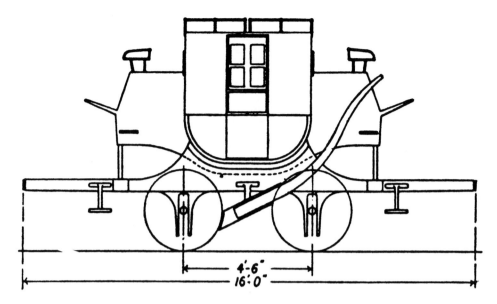

The first railway coach Experiment, *as illustrated in Samuel Smiles'* Lives of the Engineers, *vol. 3, 1862.*

carriage used on the Baltimore Ohio Railway, in 1830, bears a close and even remarkable resemblance to the Stockton and Darlington tub.[11]

Although the underframe of the original *Experiment*, supported on cast-iron wheels without springs, was made by Robert Stephenson and Company in Newcastle, the body was carefully crafted by a local coachbuilder and cost the not inconsiderable sum of £80.[12] It is hardly likely that such a superior vehicle, 'cushioned and carpeted', intended to accommodate the proprietors themselves at the opening, would have been turned out resembling a second-class 'tub'.

On the other hand, there are a number of reasons why the original *Experiment* cannot have closely resembled a stagecoach. As we have seen, Thomas Meynell referred to it as 'the *long* coach belonging to the Company'. That description is amplified in a footnote that the *Courier* of 4 October 1825 added to the account of the opening day which it reprinted from the *Durham County Advertiser*: 'It is fitted up on the principle of what are called the long coaches, the passengers sitting face to face along the sides of it.' This accords with details contained in the coachmaker's bill: 'One coach body fit up with a door at each end, glass frames to the windows, a table and seats for the inside, top seats and steps.'[13] As Tomlinson remarks: 'A carriage with a table between the two seats must have been of a fairly good width, certainly not less than 6 feet, and would consequently have the wheels beneath the frame.' These descriptions could not apply to a conventional stagecoach, but neither do they accord with Smiles' 'slab-sided box'. It also seems that the type was already known in principle if not in practice, so that the original *Experiment* may not have been novel in this respect. It is significant that the coach as depicted in John Dobbin's painting, with its picture windows, is nearer this type than either of the other two models. This version, therefore, is our preferred model for the original *Experiment*. That it proved unpopular with the public, despite its first-class status, was probably because, being unsprung, it gave them a rough ride.

Experiment, *as illustrated in Theodore West's* An Outline of the Growth of the Locomotive Engine, and a few Early Railway Carriages, *1885.*

After 'being driven off the lower part of the line' to work on the section between Darlington and Brusselton, sadly this celebrated vehicle met a rather ignominious end. It was described by Timothy Hackworth's son, John Wesley Hackworth, in a speech at Shildon on 9 November 1875:

> ... it was used on the railway for about 15 months, and was then placed at the foot of Brusselton incline, where it did service as a workmen's cabin as many old railway coaches have done since. On one occasion, when some Stockton [*sic*] men were up there, they could not get a return load, and finding the cabin comfortable, decided to pass the night there. So they got an armful of straw and laid down in front of a good fire. During the night a cinder set the straw on fire, the coach was burnt, and the men were glad to escape with their lives. Thus ended the first railway passenger carriage.[14]

It is significant that, according to this account, the first *Experiment* met its fate about the same time that the eponymous, improved, stagecoach version was introduced.

Revenue

The S&DR derived its necessary revenue from three main sources: 1. Charges for goods transported over its tracks; 2. Tolls levied on private operators; 3. Fares paid by passengers.

In this respect, while the company was granted a remarkable degree of latitude in the building of its line, the Enabling Acts spelled out a number of controls on its operation. In particular, and bearing in mind that this was a private concern answerable to its shareholders and not a public utility, Parliament imposed limits on the rates that the company could charge its customers. No doubt the legislators considered this to be a necessary and prudent precaution at a time when the absence of competition might otherwise have led to exorbitant rates.

1. **Tonnages**. Initially, and as enshrined in the First S&DR Act of 1821, 'passage upon the Railway (is) to be free upon payment of a tonnage'. The implications of this simple statement are far-reaching. It envisages owners of wagons other than the S&DR making use of the tracks which that company provided. Indeed, in its early years the S&DR's principal income was in the form of 'tonnages', so called because they were charges for the transport of goods calculated in units of 1 ton per mile. This method of charging was irrespective of the ownership of the wagons. As an example, in the 1821 Act, under the heading *Rates of Tonnage*, maximum rates were fixed for a variety of goods.[15] 'Limestone material; coal, coke etc.; Ironstone – 4*d* per ton-mile. Lead in pipe etc., 6*d* per ton-mile'. There were two significant exceptions to the limits, where it was acknowledged that the company had incurred unusual expense in providing the service. First, where goods 'shall pass the inclined planes' the company was at liberty to charge 12*d* per ton. The concession was renewed in the Second S&DR Act of 1823, and there justified on the grounds that whereas only one inclined plane was envisaged in the First Act of 1821, based on Overton's route, Stephenson's shorter route, for which authorisation was being sought, involved four inclined planes. They were also excepted in a clause in the Third S&DR (Consolidation) Act of 1849, which related to tolls of short distances of less than 6 miles, where the S&DR 'might demand the same rates as for six miles'. The second concession, after the passage of the Middlesbrough Extension Act in 1831, allowed the S&DR to charge an extra 2*d* per ton on goods that crossed the bridge over the River Tees.

In the Consolidation Act of 1849 the limits on the transport of iron ore were greatly reduced, to 2*d* per ton-mile, and a number of other commodities typical of the economy of that era were added to the list: sugar, grain, corn, flour, hides, hemp, timber, deals, metal (except iron), nails, anvils, and chains, all at 2¾*d* per ton-mile. For other commodities including coals, coke, culm, cinders, cotton and other wools, drugs and manufactured goods there were different rates ranging from 1¾*d* to 3½*d* per ton-mile, according to the distance carried, typically over or under 10 miles.

The continuing importance of livestock to the overall as well as the rural economy is exemplified in the list of animals for which, again, differential rates were charged according to distance travelled and numbers conveyed: 'Horse, mule, ass or other draught beast, ox, cow, bull or other neat cattle, calf, pig, sheep, lamb or other small animal.' Special rates for coal (together with coke, culm and cinders) for *export* were repeated, but greatly reduced to ¾*d* per ton-mile.

By 1849, of course, locomotives had taken over from horses as the principal motive power and it is significant that this Act rules out any extra charge 'for the use of the engines'. But the company was permitted to charge for loading and unloading and for the provision of covers!

2. **Tolls**. A second consequence of the appearance on the stage of entrepreneurs at the S&DR's expense was the introduction of tolls on private operators who made use of the company's line. This was simply an extension of a practice envisaged in the Second S&DR Act of 1823. The owners of individual road-running *passenger* vehicles, suitably adapted,

were permitted to make use of the railway track on payment of a toll – 'for every Coach, Chariot, Chaise, Car, Gig, Landau, Waggon, Cart or other carriage which shall be drawn or used on the Railway for the carriage of Passengers or small Packages or Parcels, such sum for each and every one, not exceeding 6d'. This provision is instructive because it is clear that in the early days the S&DR anticipated that passengers could be conveyed in a variety of privately owned, contemporary types of passenger vehicles, and not only in company-owned carriages.

3. **Fares**. Although the development of the S&DR's passenger service is dealt with in the next section, we may note here that it was not until the S&DR (Consolidation) Act of 1849 that any limits were imposed on fares. Passengers travelling first class paid up to 3*d* per mile, second class 2*d*, and third class 1¼*d*.

Above all, the proprietors of the S&DR were businessmen and therefore naturally concerned to determine where their chief sources of revenue lay. A 'Table of Traffic from its Opening October [*sic*] 1825 to June 30th 1840' is instructive in that it aggregates revenue under four heads: Coals (divided into Landsales and Export); Lime and Stones; Merchandise; and Coaches. In each category the annual revenue (June to June) is given, and in the first three categories the tonnages involved. These comparative figures are also instructive as indicating the increasing importance of passenger revenue – from £233 9*s* 5*d* in the first year to £9,677 15*s* 9*d* in 1840.[16]

Of course, all revenue had to be set against expenditure. For example, fares paid by passengers had to be offset in part by salaries paid to the necessary staff. Drivers were paid at the rate of 24*s* to 30*s* per week, firemen 15*s* to 18*s*, guards 24*s* to 26*s*, porters 17*s* to 22*s*, and plate-layers 18*s*. Stationmasters at the terminal stations received about £100 per annum, and those at the smaller stations and depots between £60 and £80 per annum. For all staff the hours were long, but especially for stationmasters – anything up to 15 hours a day. Their duties were very varied. In addition to his responsibility to passengers, a stationmaster could be expected to assist in loading trucks in the goods yard.[17]

PASSENGER SERVICES

There were at least four stages in the S&DR's provision of passenger services on its pioneer line, set out briefly here and expanded upon below:

1. For a brief period the company itself was directly responsible.
2. Private individuals were hired to operate the company's coaches.
3. Individual enterprises were allowed to operate their privately owned coaches on payment of a toll.
4. The company resumed direct responsibility.

1. In a footnote to the *Courier*'s account of the opening day, dated 4 October 1825, the newspaper announced: '[The *Experiment* coach] is intended to travel daily, for public accommodation, between Darlington and Stockton.' Within a week of the opening, therefore, and possibly even earlier, the company had determined to provide a passenger service. It should be borne in mind, however, that the proprietors 'were no doubt looking a long way ahead, for, at that time, there was little prospect of revenue from this branch of traffic, when one coach running between Stockton and Darlington three times a week afforded facilities, more than ample, for communication between the two towns'.[18] The earliest *record* of the

company's intention to undertake the carriage of passengers appears in a minute of a sub-committee dated 7 October 1825, which provides that a licence should be obtained from the magistrates for leave to run a coach on the line. (Strictly speaking, therefore, when the line was opened on 27 September, the company had no authority to make use of coaches.) That service was initiated with commendable speed on 10 October without further announcement, since the first poster-timetable refers to it retrospectively: 'The Company's Coach . . . which *commenced travelling* on Monday the 10th October *will continue to run* from Darlington to Stockton, and from Stockton to Darlington every day.'

Although the Darlington to Stockton run appears before the Stockton to Darlington run in that heading, it would appear from the timetable that it was Stockton and not Darlington which had the privilege of despatching the world's first public steam-hauled passenger train – at 7 o'clock on a Monday morning. The same train returned from Darlington at 3 o'clock in the afternoon. (It is just possible that the actual times on 10 October, and consequently the order of the runs, were different, although the words *continue to run* seem fairly conclusive.)

2. Very early in the operation of the passenger service a Thomas Close enterprisingly offered to run the coach at a wage of 2 guineas a week. The company accepted the offer on 14 October 1825 and advanced him £25 to purchase a horse and harness for this purpose. The company's interests were safeguarded by an agreement that contained the clause 'the first time he is seen intoxicated he will be dismissed, and the sum due to him in wages will be forfeited'. One suspects that the officers of the company were already privy to Mr Close's reputation for intemperance, hence this proviso. It appears that their suspicions were well founded, and that he soon yielded to temptation, for on 17 March 1826 *Experiment* was leased to Richard Pickersgill 'who took over the conveyance of passengers for the Company for £200 p.a. from 1st April 1826', although Close appears to have been retained as one of the drivers. (According to Samuel Smiles it was an anonymous driver of *Experiment* who invented railway-carriage lighting: 'On a dark winter night, having compassion on his passengers, he would buy a penny candle, and place it lighted among them on the table.')[19] Richard Pickersgill had formerly been an employee and agent of the Company. The first poster-timetable referred to above had announced that Pickersgill, at his office in Commercial Street, Darlington, and a Mr Tully of Stockton, 'will for the present receive any parcels and *book passengers*'.

The replacement stagecoach-style *Experiment* having proved popular with the public, the company therefore commissioned a second coach, named *Express*, on the same lines but 'more comfortably fitted up than the former one, being lined with cloth', to provide a faster and improved service.[20] Both coaches were leased to Pickersgill and both were timetabled from the Fleece Inn, Stockton.[21]

The contract with Pickersgill marked a radical changeover to private enterprise in the provision of passenger services. Shortly after, certain innkeepers of Darlington and Stockton, not to be outdone where there was a prospect of profit, gave notice of their intention to run rail-coaches in opposition to Pickersgill. On 20 May the company responded by cancelling the agreement that furnished Pickersgill with a monopoly of the passenger service, paying him compensation in lieu. On 2 June by-laws were passed 'for the better ordering of conveyance for passengers'. Henceforth every driver was required to deliver at the company's cottage, Durham Lane, Darlington, or at the weigh-house, Stockton, a ticket giving the name of the coach, its destination, and other particulars.

3. Given the S&DR's characteristic preference for contracting out as many services as possible where it was economically prudent to do so, it was inevitable that the passenger service would sooner or later be handed over to private entrepreneurs. The change inaugurated a comparatively long-lasting interregnum of eight years, during which private

coach owners were permitted to operate them on the S&DR's tracks on payment of a toll to the company of 3*d* per mile for each coach. One consequence of increased competition between rival railway-coach operators, and between them and the stagecoach proprietors, was an increase in the average speed of journeys by railway-coach. This soon surpassed the 10 miles an hour which was the rate of mail-coach travel and considered very fast at the time – 'the enormous speed of stage-coaches was blamed for the apoplexy of some of their passengers'![22] The speeds that had so alarmed the pundits on the occasion of the inaugural run of the S&DR were becoming commonplace.

The superior performance of the railway-coach was due to the fact that the rolling resistance of a smooth wheel on a smooth rail was very much less than that of a stagecoach wheel on a contemporary road surface. The comparison was noted by the contemporary observer Adamson in his *Sketches of our information as to rail-roads*:

> These coaches are drawn by a single horse, and yet carry six passengers inside, and from fifteen to twenty outside, besides a due proportion of luggage, and yet run at the rate of ten miles an hour. The above seems an enormous load for one horse to run with, and at such speed; and yet to look at the animal, it appears to make scarcely any exertion, certainly not so much as a horse in a gig. It is only occasionally that he gives the vehicle a pull; at other times, even in ascending from Stockton to Darlington, the traces seemed to hang quite loose; and by far the greatest exertion appeared to consist in keeping up his own motion.

One of the railway-coach proprietors who took advantage of the more liberal regime was Richard Scott, landlord of the King's Head Inn, Northgate, Darlington, who operated the coach *Reliance*. The history of the King's Head is interesting for the light it sheds on some of the changes brought about by the coming of the railways. The inn had been in existence before 1661. From 1770 it was a coaching stop for the London–Newcastle post-coach. The last turnpike coach ran in 1852, a bare twenty-five years after the advent of the railways. Perhaps Richard Scott had seen the writing on the wall when he invested in the *Reliance*. As L.T.C. Rolt has pointed out, it was only the work of the great road engineers Telford and Macadam that made possible a reliable, all-year-round and nationwide network of coach services. 'Hence it was a young and virile industry which had scarcely reached the peak of its development when the unexpected iron bolt fell out of a clear blue sky.'[23]

In addition to Richard Scott, the proprietors of the 'elegant new railway coach *The Union*' advertised 'rapid, safe and cheap transport' on the S&DR, to commence on 16 October 1926. In one direction the coach departed from the Black Lion Hotel and the New Inn, Stockton,[24] and in the other from the Black Swan in Parkgate, Darlington, near the start of the Croft branch. There was an intermediate stop at the New Inn, Yarm. Passengers and parcels were booked, and timetables kept, at all the inns mentioned, and in addition at the Talbot Inn, Darlington (on the corner of High Row and Post House Wynd), and in Yarm at the Union Inn. The Black Lion Hotel in Stockton was also the base for a rival operator, Martha Howson, who ran two coaches named *Defence* and *Defiance*.

Initially, there were no stations or waiting-rooms on the line. All the inns named as picking up points for the coaches were near to the railway, although not all were by the lineside. Those that were included inns at Aycliffe Lane, Urlay Nook, Goosepool, the Lord Nelson at Potato Hall, the Fighting Cocks at Middleton St George, and the Railway Tavern at Stockton. In such cases it is unclear how the coaches were transferred from rail to road and from road to rail, that is from one set of wheels to the other.

Early private railway coach The Union.

Within fifteen months of the opening of the railway, seven regular coaches were operating between Darlington and Stockton. In the accounting year July 1826 to June 1827 they covered 45,450 miles and carried 30,000 passengers, but only earned a modest £503 in fares. Not surprisingly, these first *railway* passenger coaches were given names reminiscent of their road-running counterparts. An article on early rail transport which appeared in an issue of *The Graphic* dated 13 October 1888 related the following anecdote:

> In designing the carriage for the early railways, the coach was the type naturally selected. In many cases each carriage had a distinctive name, and a story is told of the consternation of the officials on a Northern Railway, who found that they had unintentionally placed the 'Waterloo' carriage at the disposal of a French General during his visit to the district.

Because the railway-coaches were run by six different proprietors, a great deal of rivalry resulted as each sought to obtain operating advantages over his competitors on what was effectively a 'public highway'. The line was single and devoid of signals or any other form of train control, but passing loops were provided at quarter mile intervals. The order of precedence as to right of way was complicated and the regulations were difficult to enforce:

Locomotive-hauled coaches and wagons had the first priority and all horse-drawn coaches and wagons had to give way for them, reversing if necessary to the nearest passing loop.

On gradients, descending locomotive-hauled trains of wagons had to give way to ascending trains of locomotive-hauled wagons and occupy the passing loop.

If they met on gradients between passing places the ascending train had to reverse back and occupy the nearest passing loop.

A locomotive-hauled train of empty wagons always gave way to a locomotive-hauled train of loaded wagons.

Next in the order of precedence were horse-drawn trains of loaded wagons – horse-drawn coaches had to occupy a passing loop if approached by any of the previous categories, reversing if necessary.

On the other hand, if a faster horse-drawn coach overtook a train of wagons (whether locomotive-hauled or horse-drawn) the latter had to stop beyond the switches to allow the coach to pass into the loop and out at the other end without stopping, ahead of it. (Alternatively the train of locomotive-hauled wagons would enter the loop to enable the coach to pass it on the straight line.)

The greatest difficulty was experienced when two horse-drawn coaches were travelling towards each other. On approaching a passing loop a passenger coach was supposed to wait there until another arrived from the opposite direction. Such rules were made to be broken, and in practice coaches often overran the loops in order to gain time. Inevitably, this resulted in two coaches meeting head-to-head on the intervening single track. Disputes then arose as to which should go back, frequently degenerating into fist-fights, and sometimes even involving the passengers. Stagecoach drivers themselves had never been noted for their civility! Timothy Hackworth was provoked into exclaiming to the management committee, 'Gentlemen, I only wish you to know that it would make you cry to see how they knock each other's brains out.'[25] In an attempt to resolve the problem, posts were placed midway between the passing loops with the instruction that 'first past the post' should have precedence (hence the expression), the other being obliged to retreat to the nearest passing loop. Needless to say, this did not resolve the problem, as desperate efforts were made by the drivers of opposing coaches to reach the post first. 'A permanent state of war prevailed on the line which not infrequently reduced traffic movement to chaos.' Certainly, the railway-coach traffic was a happy-go-lucky business during these early years.

L.T.C. Rolt cites several examples of the consequences of a wilful disregard of the 'first past the post' rule, of which the following are typical:

Two horse-drivers refused to allow a steam train to pass them by entering the loops and forced it to follow them for 4 miles. On the same day another driver shunted some wagons so violently that his horse was pitched out of the dandy cart and fell down Myers Flat embankment.

Ralph Hill met a locomotive train and absolutely refused to give way although the engine had passed the midway post. He was told not to bring his horse on the railway again and was summoned to appear before the magistrates.

Two horse-drivers, William Ogle and George Hodgson, left Shildon in such a roaring state of drunkenness that Ogle continued to drive his horse at a gallop regardless of the fact that the dandy cart had jumped the rails and was tearing up the track. They forced another horse-driver whom they met to go back into a loop and overturned his empty

Daniel Adamson's railway coach-house, Shildon.

wagons. Finally, this precious pair met a steam train and, refusing to give way, tore up a rail, threatening to throw the engine off the line.

These were only some of the difficulties, trials and tribulations with which John Graham, the S&DR's first traffic manager (and therefore the world's first traffic superintendent) had to contend. John Graham was head overlooker at Hetton Colliery when he was recommended for the new post by George Stephenson. We owe much of our knowledge of the practices of the S&DR in its early years to the detailed diaries and reports that Graham prepared. From these it is clear that the proprietors were persuaded to sanction the progressive lengthening of the passing loops so that by 1830 it was reported that there was a 'gradual approximation to a double line of railway'. Soon afterwards the line was completely doubled; had that been the case from the beginning many of the logistical problems encountered by Graham as traffic manager would never have occurred. It is true to say, however, that the doubling of the line arose as much from the continuing expansion of traffic following the opening of the Middlesbrough extension as to the vexatious operational difficulties that were being experienced.

One very tangible reminder exists of the early railway-coaching days. Daniel Adamson was the landlord of the Grey Mare (or Grey Horse) Inn at Shildon, which was patronised by packhorse men and wagoners travelling through with merchandise and coal. In due course it became a house of call for railwaymen also. (Part of this inn is incorporated in the present Surtees Arms, which replaced it.) In 1827 Adamson, who was also a local coal 'leader', pioneered a passenger service with his railway-coach *Perseverance* between Shildon and Darlington from the Mason's Arms. In 1831 the Surtees Railway was opened from New Shildon Junction to Shildon Lodge (Datton) Colliery, which was owned by the Surtees family and conveniently adjacent to the Grey Mare Inn. Adamson accordingly had a coach-house built alongside the railway, to serve as both a station and a shed. From here the coach ran along the Surtees Railway, and thence over the S&DR to Darlington. The coach-house, on the corner of Byerley Road and Main Street, has been preserved.

4. The disorganised state of the passenger service increasingly became a major concern of the company. It took a first step towards resolving the confusion when on 22 January 1830 it issued a regular timetable of all coach operations, which came into effect on 1 February.[26] (Richard Pickersgill and Richard Scott are specifically named.) On 2 November 1832 the company issued a 'Statement of the number of coaches which have travelled or plied for hire on the S&DR during the last twelve months'. The list of proprietors included Richard Pickersgill and Company; Richard Scott and Company; Ludley and Buckten's; Messrs Adamson; Messrs Wastell; and Messrs Harris. A few months later, mindful no doubt of the coaches' increasing profitability, the company resolved to run the passenger services itself from 7 September 1833. This involved 'buying out' all the independent coach proprietors. They accordingly yielded up their franchises and sold their rail-running coaches to the company for the sum of £316 17s 8d.[27] This radical resumption of direct control was bound up with a long-running, thorny question that is dealt with in the section 'On Trial': Locomotive Faults and Failures below.

Goods Wagons

There was nothing in the First Act to prevent entrepreneurs contracting with the owners of goods to provide wagons as and when required, thus relieving them of a considerable capital outlay. Since, in effect, these private operators were making a profit at the S&DR's expense, the company responded by increasing its own complement of wagons so that it could directly provide the kind of service that has since become the norm throughout the history of railway goods traffic, where it is the *railway company* itself that provides the rolling stock. In any event, it was necessary for the S&DR to be able to distinguish between 'company' and 'private' wagons, and to identify the owners of the latter so that they could be charged the appropriate tonnage on their contents. The 1821 Act accordingly made provision for the 'owners of waggons or other carriages [to] cause his or her name, Place of Abode, and the number of his waggon or other carriage [to be] painted in large white letters and figures on a black ground three inches high at least, on some conspicuous part of the outside of such waggon or other carriage'.

In 1825 the only rolling stock available to the company were 150 chauldron wagons, most of which had been used for horse-hauled coal traffic on colliery wagonways. Indeed, some had been brought from the Hetton Colliery Railway for use in constructing the line. (It will be recalled that, apart from the coach *Experiment*, the inaugural train was entirely composed of chauldron wagons for the conveyance of coal and passengers alike.) These wagons had a tare weight of 28–32 cwt and a capacity of about 2 tons – a 'London chauldron'. They were not suitable for the Stockton staithes for two reasons. First, they fell short of the statutory capacity (53 cwt – a 'Newcastle chauldron') by some 13 cwt,[28] and second, they could not be discharged straight into the holds of the ships as they had end boards and not bottom boards. As early as 21 October 1825 Thomas Storey (the assistant resident engineer) wrote to the management committee complaining that in constructional terms they were 'nearly all as bad a set of waggons as were ever turned out on a Railway' because of inferior wood and flimsy sides.[29] In their existing state they could not be expected to last more than two years.[30]

Chauldron wagon preserved at the National Railway Museum.

Within three days of the opening, the committee considered adopting the so-called Wear system in which the coal was carried in boxes each containing not a Newcastle chauldron of 53 cwt, but a London chauldron of 40 cwt. Although George Stephenson was instructed to furnish the required dimensions, it was decided instead to alter most of the existing wagons by adding side boards at the top so as to increase the capacity to 53 cwt. At the same time, all the wagons were adapted with bottom boards in place of end boards, which were released by retaining pins at each side. Discharge through the bottom of these chauldron wagons was aided by their inward sloping sides. These utilitarian vehicles were of very basic but sturdy construction, with crude wooden square-section buffers, and primitive brakes consisting of wooden blocks bearing on the front wheels, operated by a long lever.

Quite apart from the question of capacity, it soon became apparent that, suffering as they did from constant wheel breakages and consequent derailments (mainly because of the use of fixed axles), chauldron wagons were quite unsuitable for transporting general merchandise, which was either damaged or flung out onto the lineside. In any event there was a shortage of wagons to cope with the demands of the traffic, giving rise to numerous complaints from customers. There was therefore an urgent need for new and improved wagons. Thomas Shaw Brandreth, a Liverpool barrister, patented an improved wagon on 8 November 1825 and some of these were purchased by the S&DR. After experiments with Brandreth wagons in September 1826, Robert Stephenson and Company introduced a new spring-mounted wagon utilising 'anti-friction rollers', that is outside bearings in axle boxes. On 27 November 1928 the S&DR took delivery of the first of this new generation of 'improved' wagons, each carrying about 4 tons,[31] which was to become the prototype of the standard British goods wagon.

Dandy cart preserved at the National Railway Museum.

In the early years of the S&DR there were problems with horses when hauling freight, particularly on the falling gradients of the line, where, despite the best efforts of the brakesmen, the vehicles had a tendency to overtake the animal, which had to try to race ahead of its train of four chauldron wagons. This disadvantage was overcome in the summer of 1828 with the introduction of the dandy cart, 'a simple and ingenious contrivance which economised their [horses'] powers to the extent of one-third'. Although Thomas Brandreth claimed credit for this idea, it was originally suggested by George Stephenson in 1826 for the Canterbury and Whitstable Railway, but it had not been adopted there.[32] The dandy cart was a special, low vehicle – 'a homely waggon with low wheels' – attached to the rear of the coal wagons and equipped with a box of hay and a bucket of water, in which the horse was able to ride over the downhill sections, or 'runs', thus enabling it to rest, eat some of its fodder, and be refreshed. Thus the horse rode in the dandy cart from Shildon to the foot of Simpasture (3 miles); from Aycliffe Lane to Darlington Junction (4½ miles); from Fighting Cocks to Goosepool (2 miles) and finally from Urlay Nook to Stockton (4 miles). The horse 'dismounted' to pull the loaded wagons on the intermediate stretches and, of course, returning with empty trains had to pull all the way.[33]

It is said that the horses soon got used to the new method of working. Immediately the driver unhooked the traces,

> he would allow the waggons to pass him and then trotting after the train, leap onto the low dandy of his own accord, and he performed the feat not only without urging, but on the contrary with so much eagerness as to render it difficult to keep him off, although the carriage was two feet from the ground, and the progressive rate nearly 5 mph.

A basket of hay was suspended from the dandy, and therefore the only wonderful part of the ceremony was its performance.[34]

Once aboard, the horse rested continuously on its haunches against the contingency of a sudden stop. Like the chauldron wagons, the dandy cart had a tiller or brake to assist in controlling its speed down the runs. In some illustrations, the presence of a fire-bucket indicates that dandy carts were operated during the hours of darkness.

The introduction of the dandy cart enabled the company to get an average of 240 miles out of each horse per week, compared with the earlier 174 miles – an increase of 40 per cent – and, assuming the carts were working efficiently, to narrow the gap of advantage enjoyed by the locomotives. To overcome an initial reluctance on the part of the coal leaders to adopt the dandy cart, in November 1828 the company ordered that 'every horse-leader of coals shall immediately provide himself with trucks for carrying the horse down the descending part of the line, as no encouragement of any description will be given to any leader who does not provide this accommodation'. That was the stick. However, bearing in mind that the carts cost the horse-leaders an expensive £6 each, the company agreed to pay an extra 1d per ton of coal carried in trains equipped with dandy carts. That was the carrot! Although dandy carts were only employed on the main line for a short while, they persisted elsewhere on the system until 1856.[35]

'ON TRIAL': LOCOMOTIVE FAULTS AND FAILURES

For some days after the opening the only engine available to carry the burden of traffic was *Locomotion*, until the second engine, *Hope*, was delivered early in November 1825, much later than promised. The proprietors must have been satisfied with *Locomotion*'s performance on the inaugural run for shortly after they placed an order for a further two sister engines, No. 3 *Black Diamond* and No. 4 *Diligence*, which were delivered in April and May 1826 respectively. The proprietors' satisfaction did not last long, however, as shortly after they took possession of No. 2 *Hope* the company minuted: 'This Committee feels very much dissatisfied with the manner in which Messrs Stephenson and Co, have delivered the last Locomotive Engine on account of its very imperfect state, the Smiths having been employed a whole week before it could be got to work.'[36]

This was but one example of a whole series of mechanical deficiencies that put *Locomotion* and her sister engines out of action for months at a time. Ironically, *Locomotion* herself broke a wheel within days of the opening and was out of commission until 12/13 October. It was probably this and similar incidents that were being referred to when, only a week after the committee had complained about the incomplete condition of *Hope* on arrival, it was further minuted:

> Disappointments by the breaking of various parts of the Locomotive Engines having repeatedly occurred especially by the breaking of the wheels – resolved that Robt. Stephenson and Co. be requested that in any Engine they furnish us with not to send any Engines with new and experimental apparatus, but such fitting up as hath been tried and approved already, and that as the disadvantage arising from not having duplicate wheels for the engines are very great, Robert Stephenson and Co. are requested to send some spare wheels fit on the axles which it is desired may be always of the same pattern not only for the Engines but for the Waggons.[37]

The management committee's frustration at this time must have been extreme for only one week later they adopted a further resolution:

> Innumerable accidents and inconveniences having arisen from some defects in the Locomotive Engines and this Committee considering the great expenses they have incurred in alterations and repairs . . . direct the Clerk to request [Robert Stephenson and Company] will *protemporare* place a Smith or person sufficiently acquainted with Locomotive engines at Brusselton or elsewhere in order to superintend the alterations and repairs which attach to the said Engines on account of [them] not being perfect and complete when set to work.[38]

As a case in point, and despite the fame now attached to the first steam engine ever to run on a public passenger-carrying railway, it must be said that *Locomotion* was prone to accidents, the result of these mechanical defects plus the fact that she was notoriously difficult to drive. George Graham, son of John Graham, who was one of *Locomotion*'s drivers, commented that not one man in three had the ability to drive her and her four sister engines. This is confirmed in a letter from Robert Stephenson to Michael Longridge dated 1 January 1828: 'Mr. Jos. Pease writes my father that in their present complicated state they cannot be managed by "fools", therefore they must undergo some alteration or amendment. It is very true that the locomotive engine . . . may be shaken to pieces; but such accidents are in a great measure under the control of the enginemen, which are, by the by, not the most manageable class of beings. They perhaps want improvement as much as the engines.'

The drivers' difficulties were compounded by the fact that, since fuel economy was not a priority, for much of the early life of the S&DR fuel was the driver's responsibility. He was paid at the rate of a farthing per ton per mile, out of which he, in turn, had to pay his fireman or firemen, and to supply fuel, oil and firebars. Although from the company's point of view this practice proved effective, it led to many abuses. There was a constant incentive to break the speed limit of 8 mph, which was imposed in an attempt to reduce wheel breakages. Speeding, however, was common, especially if the drivers were delayed by a breakdown or by other traffic. The record for speeding was gained by a driver named Charles Tennison, who covered the 45 miles from Shildon to Stockton and back, including stops, in four and a half hours: 'For this feat he was instantly dismissed.' Although drivers caught speeding were heavily fined, it proved impossible to enforce the speed limit in practice. It was also customary in these early days for members of the public to 'hitch a lift' on the engine and tender, with the connivance of the drivers. It is clear that, initially at least, safety was *not* a priority.

In March 1828 *Hope*'s boiler blew up, killing the driver John Gillespie. This was the world's first recorded fatality for a railway in operation. While taking on water at Aycliffe Lane on 1 July 1828, *Locomotion*'s flue burst, killing the driver, John Cree,[39] and maiming the water pumper, Edward Turnbull. Timothy Hackworth discovered that these boiler explosions resulted from the drivers' suicidal habit of tying down the Stephenson safety-valve arms with cord once they were out of sight of the Shildon sheds to obviate the nuisance of intermittent jets of steam escaping from the valve caused by the unevenness of the track. This revelation prompted Hackworth to introduce his spring safety-valve, which effectively put a stop to this dangerous practice.

Since John Cree was for long the regular driver of No. 1, it was probably *Locomotion* that was referred to as 'Cree's engine' when it overturned at New Shildon on another occasion, requiring repair. The operational difficulties were exacerbated by the fact that, in common with her sister engines, neither *Locomotion* nor its tender were fitted with brakes. The only way to stop was to reverse the engine. As there was no reversing gear, the eccentric rods had to be lifted and the valves worked by hand – a complicated manoeuvre. George Graham's comment that this was 'no easy matter when running at 12 miles an hour or more' was clearly an understatement. At times, these operations made the task of stopping the engine hazardous in the extreme, especially when hauling a heavy train of chauldron wagons on some of the descending gradients between Shildon and Stockton.

Mishaps were commonplace. In 1833 *Locomotion* ran into a donkey near Goosepool and the fireman lost a foot. In November 1837 it overturned at Aycliffe Lane because of careless work on the track, causing damage estimated at £22 16s 10d. In 1838 it ran into some wagons at Goosepool and sustained further damage. As a correspondent of the *Railway Times* of 2 October 1875 wrote, with some justification: 'It is impossible to look at No. 1 and not to admire the pluck of the men who rode this spider-legged patriarch at the rate of 15 mph.'

The inevitable consequence of these mechanical breakdowns, and the accidents already mentioned, was that, for long periods of time, the company was obliged to resort to horse power as the sole means of hauling its heavy trains of chauldron wagons. (So unreliable were the first generation of locomotives that the company at one stage even seriously considered abandoning steam permanently in favour of horses and gravity.) Even if the steam engines had been reliable machines, they would have had difficulty hauling all the freight traffic that had been generated. In the year ending 30 June 1827, for example, this totalled 100,000 tons. In any case, for many years after the opening horse power had the monopoly of hauling the passenger coaches. (For example, in the handbill announcing the beginning of the service provided by the new coach *The Union* on 16 October 1826, it is shown drawn along a railway track by a high-spirited horse and not a locomotive.) Steam engines would only have been economically viable when hauling passenger coaches if they were coupled together in trains, analogous to the 'freight trains' of chauldron wagons. Rightly or wrongly, the public believed that this unfamiliar practice would be dangerous. It is largely for this reason that single horse-drawn, rather than locomotive-hauled coaches in preference to locomotives persisted on the railway for almost a decade.

Even when the company's steam locomotives were repaired and in working order, hauling freight, they were under another severe handicap quite apart from their mechanical unreliability. Since they were obliged to share track cluttered with horse-drawn trains of chauldron wagons and coaches of the 'by-traders', this played havoc with the schedules. Those proprietors who initially had doubts about the steam locomotive as a quick, reliable means of haulage must have had their worst fears confirmed. Nevertheless, the company itself was largely to blame for pursuing a mistaken policy of 'joint use', since in practice this led to the ascendancy of 'horse power' over 'steam power' for the greater part of the years 1825 to 1833. As Robert Stephenson wrote:

> scarcely a single journey is performed by [the] engines without being interrupted by the horses or other trains of carriages passing in a contrary direction there being only a single line of road with passing places. At each end of the distance traversed by the engines great delay is occasioned from the irregular supply of carriages which from the nature of the trade and other local circumstances it is impossible to avoid.[40]

It is little wonder that in such a milieu the steam locomotive could not fully demonstrate its superiority.

THE EARLY YEARS: PRIVATE RISK FOR PUBLIC SERVICE

Timothy Hackworth's locomotive the Royal George. *(From* The Engineer*)*

'REPRIEVE': LOCOMOTIVE IMPROVEMENTS

Matters came to a head in the summer of 1827. At this time the company possessed four working locomotives of the 0–4–0 'Locomotion' class. In spite of their general unreliability, they were some 30 per cent cheaper to operate than horses, as calculated by the company's first chief engineer, Thomas Storey.[41] At the beginning of October 1827, however, one of them ran away along Stockton Quay and did much damage both to itself and to the installations. It was now patently obvious that the company's chief priority was to secure not only more locomotives, but more efficient and dependable locomotives. There is some debate whether, notwithstanding the numbers allocated to them in the company's records, the first of these 'improved' engines was No. 5 *Royal George*, designed and built by Timothy Hackworth at New Shildon, or No. 6 *Experiment*, designed by George Stephenson and built at Robert Stephenson and Company's works at Forth Street, Newcastle. In their Report, the two Prussian mining engineers describe *Experiment* as they saw her, complete, at Forth Street early in 1827. However, there is no record of her working on the S&DR before January 1828. On the other hand, *Royal George* began working in November 1827.

The 0–4–0 *Experiment* was 'in a class of its own'. It was the first to depart from the vertical cylinder convention that had prevailed since Blenkinsop and Murray's rack-railway engine. In *Experiment*, Stephenson revived Trevithick's practice of mounting the two cylinders horizontally in one end of the boiler. A complicated system of levers and

Locomotive The Globe, *at the head of Timothy Hackworth's business card.*

connecting rods transferred thrust from the pistons to the leading driving wheels.[42] It was Timothy Hackworth, however, as resident engineer of the S&DR, who undertook the formidable task of restoring the steam engine's reliability and prestige. The result was the *Royal George*, the first really successful locomotive on the S&DR.[43] The *Royal George* was the prototype for a series of classes of six-coupled goods engines which were the mainstay of the railway for many decades to come. Not for the first time, and certainly not the last, Timothy Hackworth's ingenuity and invention saved the day for the S&DR.

The rehabilitation of the locomotive was signalled in July 1828 when Hackworth was instructed by the management committee that 'the engines are in future to take all coals possible' with horses receiving the residue. This was followed in September by another directive that 'the coal owners, east of Brusselton plane, are to send their coal by locomotive engines as formerly, and if they are unwilling they must apply to the Company'.[44] 'A further act of coercion for recalcitrant colliery owners came in July 1829 with the decision of the management committee that wagon facilities should be withheld from traffic senders who refused to co-operate with the company's policy on locomotive haulage'.[45]

This decision in 1828 to phase out horse haulage for mineral traffic was to prove critical in ensuring the company's subsequent commercial success. It was the dandy cart's misfortune that this innovative device was introduced in the same year. Although its employment on the S&DR 'main line' was short-lived, it continued to give good service on branch lines, including the Black Boy branch.

Having successfully designed a locomotive that would handle goods traffic on the S&DR, Hackworth turned his attention to passenger traffic. The result was No. 9 *The Globe*, to his design, manufactured by Robert Stephenson and Company at Newcastle and delivered in 1830. *The Globe* had the distinction of being one of the first engines to have inside cylinders and a cranked axle and was the first light four-coupled locomotive intended for passenger train working on the S&DR, as well as goods. It is reported to have attained speeds of up to 50 mph in favourable circumstances. Significantly, a vignette of *The Globe* headed Hackworth's business card.[46]

By 1832 the S&DR possessed nineteen locomotives, twelve of which had been built in the short period 1831/2. Having resolved the manifold problems resulting from the unreliability of the first generation of steam engines, the time was right to substitute steam power for horse power throughout the system. It should be borne in mind that by this date Stephenson engines had been successfully and regularly hauling passenger and freight trains on the L&MR for almost three years. After the Rainhill trials, 'the battle for the locomotive had been most decisively won'.[47] In April 1833 John Graham, as traffic superintendent, was able to report that on the main line all coal traffic was now being 'led' by locomotives. The complete substitution of steam power for horse haulage on the main line was approved at a meeting of the management committee on 30 August 1833 when it was decided to extend the change-over to passenger traffic also.[48] This was also the catalyst that enabled the company to take into its own hands the direct operation of its passenger services from 7 September 1833. By the end of the year arrangements were well in hand for steam-hauled passenger and goods services between Shildon and Middlesbrough. The transition was finally accomplished in April 1834 with the inauguration of a passenger and goods service on the Middlesbrough extension.[49]

With the introduction of locomotive-hauled passenger trains a distinction was made between locomotives that were suitable for coal and merchandise work on the one hand, and passenger traffic on the other, for which two separate departments were set up under the supervision of the resident engineer. A 'Statement of Work Done and Expenses Working and Repairing Locomotive Engines from July 1st 1838 to 30th January 1839' lists twenty-seven named and numbered locomotives (including the original four from the Forth Street Works and *Royal George*), allocated to the coal and merchandise department, and five named but unnumbered locomotives allocated to the coaching department. The practice of contracting out is again apparent. A footnote reads: 'The coaching train engines were maintained by Engine Builders who were Contracted and were paid at the rate of 7d per train mile.' The statement was signed by John Graham as operating superintendent.[50]

LINESIDE

We end this chapter with an account of certain features which would have been observed by the lineside of the S&DR.

Signals

In the very early years of the S&DR, signalling was not accorded first priority and was developed on an *ad hoc* basis – drivers were expected to look after themselves and their trains. Near level-crossings, square boards with 'Signal' painted on them were fixed to posts to warn drivers to slow down and ring their bells. However, with the increase in traffic the introduction of an efficient and dependable signalling system became more pressing. A variety of methods was proposed, some of which were put into practice on a trial basis. A common feature was to mount a board on top of a pole that could be rotated. An early version peculiar to the S&DR consisted of a triangular board that signalled 'danger' when set at right angles to the track and 'clear' when set edge-on. In another version, the triangle was replaced by a red disc for 'danger' in the 'on' position and 'clear' in the 'off' position, set end-on. The most sophisticated version was developed by William Bouch, the company's engineer at the time, in which the disc was replaced by two rectangular boards, one painted red for 'danger' and the other white for 'clear', set at right angles to each other so that by rotating the pole through 90 degrees either could be seen by the driver. A lamp with corresponding red and white lenses was also mounted on the pole beneath the boards and rotated in unison. For night-time operations, when the unilluminated boards could not be seen, a crude arrangement of a brazier hoisted aloft on a gallows with rope and pulley, signifying 'danger', was used instead.[51] By 1852 semaphore signals were replacing the old signal boards for use both by day and night.

It appears that the junction on the 'main line' at Albert Hill, Darlington, was first appropriately named Croft Branch End and later York Junction. It was here that what was arguably the world's first rudimentary colour-light signal was installed. 'It was a pole fitted with a pulley from which, suspended by a chain hung a lamp 4 feet high . . . and one foot wide . . . , "caution" being enjoined when the slide was pulled half way up and a green light shown and "danger" indicated when the slide was pulled the whole way up and a red light shown.'[52]

Telegraph

It was a happy chance that the railways and the electric telegraph, invented by Charles Wheatstone (1802–75), were introduced at much the same time. (The telegraph was first applied to railways by the GWR in 1839.) Because the telegraph enabled messages to be transmitted in advance, faster than the trains themselves, one of its first applications was to enable railway clocks to be synchronised all over the land – a standard 'Railway Time' was a prerequisite for railway timetables. Without the telegraph, the 'block' system of train operation, which became universal and which depends upon the ability of signalmen to communicate with each other, could not have been introduced. The rapid spread of the telegraph system was facilitated by the expansion of railways throughout the land, since they provided a ready-made, convenient network of tracks alongside which the telegraph wires could be laid. Accordingly, on the 1861 OS map, the word 'Telegraph' is printed by the side of S&DR lines. When the single-track Middlesbrough and Guisborough Railway opened in November 1853, the electric telegraph and the block system were in use from the start.

Property Plates

These were a common feature on the S&DR, and were attached to dwellings but not operational buildings. A whole terrace warranted only a single plate. They were made up of a letter and a number, and the sequence ran from east to west across the railway network as it existed in 1856–60. Despite the remarkable persistence of S&DR dwellings, very few property plates remain *in situ*, the majority having been removed by museums and souvenir hunters. Those that do remain are mentioned in the text, as appropriate.

An S&DR milepost in situ.

Milestones

Early in its history the company settled on a simple, standard milestone consisting of a sandstone block with rounded edges about 2 ft high. Each milestone carried only the company's initials, S&DR, in a characteristic adaptation of Roman script but with serifs. There was no other inscription, not even a numeral to denote the location in relation to another important starting point. Again, milestones *in situ* are now very rare, but one does occasionally come across a relic by the lineside in remote locations.

CHAPTER 5

Real Estate and Customer Services

A *Public* Railway?

The changing nature of the company's provision of a passenger service leads us to examine the meaning of the word 'public' as applied to the S&DR in the early years of its existence. The original concept of the function of the S&DR was to provide the permanent way and to allow anyone who wished to use it to do so, according to the time-honoured principle that any highway should be open to all on payment of tolls. Indeed, in the early years, the railway was used as a public footpath, and when that was restricted it continued to be used by doctors between sunset and sunrise with the sanction of the directors, and for funeral processions 'paying an acknowledgement of 6*d* for the privilege'.[1]

'The railway was intended to be a highway for all comers, and the haulage advantages were open to the public on payment of certain fees and tolls',[2] and provided they obeyed the relatively small number of rules laid down for safe and convenient operation. According to the Act of 1823 the public were to be free 'to use with horses, cattle and carriages' the railroads formed by the company, between 7.00 a.m. and 6.00 p.m. in winter, between 6.00 a.m. and 8.00 p.m. in April and September, and between 5.00 a.m. and 10.00 p.m. in summer. In addition, it is true, the company also intended to make use of its own route for its own passenger carriage and coal trains. As we have seen, however, the first of these was soon entirely given over to private enterprise. One reason may have been because the company's single coach *Experiment* proved unpopular with the public, who preferred to use the stage-coaches which were adapted by private operators to travel on rails. The company may well have been reluctant to invest in the capital cost of constructing this type of coach *de novo* because of the financial constraints of the early years. (Some years later Parliament was to decree that any new railway or major extension must itself provide a passenger service from the beginning.)

The regulation of the carriage of coals and merchandise was a much more important consideration, since this was the *raison d'être* of the railway. Even before the opening, it became clear that, unlike the passenger service, it would be impossible to regulate the often competing operations of a number of uncontrolled carriers. Any system that allowed independent owners of trains of wagons to make use of the track on payment of a toll would have been an operational nightmare. The proprietors therefore reserved to themselves the monopoly on the carriage of the most important commodity – coal. In addition to the profits accruing from that trade, which the railway facilitated, it was essential to maximise revenue from other sources. From the beginning, therefore, the 'public', in this case independent coal owners and merchants, were invited to consign their goods for carriage in the company's own wagons.

The solution adopted by the S&DR was to make use of a class of middlemen known as 'coal leaders'. As the name suggests, each coal leader was in charge of a train of four chauldron wagons. While the company owned the wagons and transacted the business, the coal leaders were paid according to a formula that took account of not only the amount of work done but also such factors as the speed and efficiency of the operation. For example, the coal leaders were paid a premium if they made use of dandy carts, after their introduction in 1828, because this speeded up operations on the downhill sections and so maximised the company's profits.

It is important to emphasise that even influential proprietors such as members of the Pease family, who owned a substantial proportion of the shares in the company, were nevertheless on a par with any other coal owner or merchant who made use of the railway. They enjoyed no preferential treatment or any discount on the company's charges.

A huge nuisance had been caused by the horse-drawn passenger traffic, under separate and independent control, hindering any sort of regular steam-hauled traffic. With the introduction of more reliable locomotives, and the buying out of the passenger franchises in 1833, it became possible to rationalise and unify the passenger and goods services. A poster advertising the railway's winter operations commencing on 1 October 1840, included the times of separate passenger and merchandise trains. Only fifteen years after the opening, the timetables for three passenger services are given – between St Helen's Auckland and Darlington; between Darlington and Stockton (first-class trains); and between Stockton and Middlesbrough. Significantly, between Darlington and Stockton the times of three 'Merchandise Trains' in each direction are also given. By 1840, therefore, small horse-drawn trains of only four wagons, with a coal leader at the head, had been replaced by much larger, heavier trains hauled by efficient steam locomotives, with a crew of only three men employed directly by the company. The question of motive power had been unequivocally resolved in favour of steam, much to the advantage of proprietors and public alike.

PEASE FAMILY HOMES

Away from the line, a number of buildings in Darlington have associations with the S&DR in its early days, and in particular with its principal founders, the Quaker Pease family. The Peases were first established in Darlington in the person of Edward Pease (1711–85), woolcomber, who as a youth came to work in the woollen mill of his maternal uncle, Thomas Coulthard. Edward must have gained the confidence of his uncle, for in 1744 Thomas Coulthard retired from the business, handing it over to his nephew. Edward's son Joseph (1737–1808), to whom he in turn passed on the business, bought a house called The Bull, in the eponymous Bull Wynd (in the centre of the borough), on the occasion of his marriage to Mary Richardson in 1763. The name of the house was changed to The Grove, and it was here that our Edward Pease was born on 31 May 1767, the middle one of five children.

Up to the age of twelve, Edward was educated locally, and then sent to boarding school in Leeds. When he returned home he entered the family business – wool merchants and weavers – in Darlington. When he married Rachel Whitwell of Kendal in 1796 he brought her to live in his parents' house, The Grove. The house still exists, and a row of premises to the south of the building has been demolished to give an uninterrupted view from Houndgate. The area thus cleared has been sensitively converted into a forecourt, and the house itself has been carefully restored and renamed Pease House.[3]

In 1798 Edward and Rachel moved to a larger house, with gardens extending down to the River Skerne, at what is now 146 Northgate. It was here that the historic meeting with George Stephenson took place, and where the engineer later converted two rooms for use as

The birthplace of Edward Pease – The Grove, Bull Wynd, Darlington.

Edward Pease's home, 146 Northgate, Darlington.

his offices during the construction of the line. This arrangement proved to be a convenient one for the two pioneers: 'In top boots and breeches George Stephenson and John Dixon would work all day long from dawn to dusk surveying the new line, and Stephenson would constantly drop in when the day's work was done to discuss with [Edward Pease] the Railway and various matters.'[4] The present building, opposite the former Technical College, is not the original but a later one on the same site. An even later Art Deco oriel window, totally out of keeping with the rest of the facade, projects from the first floor. Two ceramic plaques were placed on it in 1942, above and below the window proper. The one above reads 'The First Public Railway was Inaugurated Here', and the one below: 'The House Where Edward Pease resided 1820'.

After the death of his wife on 18 November 1833, Edward Pease withdrew from public life altogether, and passed over the management of his business affairs, including his involvement

George Stephenson instructing Mary and Elizabeth, daughters of Edward Pease, in the art of embroidery in their home in Northgate, Darlington, 1823. (Engraving by W. Thomas for the Illustrated London News, *4 January 1862. From a painting by Alfred Rankley (1819–72).*

with the S&DR, to his son Joseph. From 1833 until his death at No. 146 on 31 July 1858, aged ninety-one, he devoted his time increasingly to the care and enjoyment of his house and gardens.

One celebrated incident that took place in Edward Pease's home is recounted by Samuel Smiles. After the day's work was over, he writes, George would drop in to talk over the progress of the survey, and discuss various matters connected with the railway:

> Mr. Pease's daughters were usually present; and on one occasion, finding the young ladies learning the art of embroidery, he volunteered to instruct them. 'I know all about it', said he; 'and you will wonder how I learnt it. I will tell you. When I was a brakesman at Killingworth, I learnt the art of embroidery while working the pitmen's buttonholes by the engine fire at nights.' He was never ashamed, but on the contrary rather proud, of reminding his friends of these humble pursuits of his early life. Mr. Pease's family were greatly pleased with his conversation, which was always amusing and constructive; full of all sorts of experience, gathered in the oddest and most out-of-the-way places.[5]

This scene was the subject of a painting by Alfred Rankley (1819–72), *George Stephenson at Darlington in 1823*, which was exhibited at the Royal Academy[6] and shortly after engraved by W. Thomas for an issue of the *Illustrated London News* on 1 January 1862.

The accompanying text is instructive because it indicates the esteem in which George Stephenson was held at the time:

> This little incident in the domestic life of a great inventor and great public benefactor is to us of extreme interest. Mr Samuel Smiles tells the story in his life of the illustrious founder of our Railway system. There is an amiable simplicity in the whole of this family group and of the expression of the faces which suggests the delightful calm enjoyed in the society of those he loved, in snatched moments of leisure, by the hard-thinking and hard-working engineer. His face, which is, we believe, a correct portraiture, beams with that brightness of intelligence, coolheadedness, and orderly thought for which he was so eminently distinguished.[7]

To add authenticity to the scene, Edward Pease is shown seated at a table on which there is a diagram of a Killingworth engine, and holding a plan of the S&DR.

There is circumstantial evidence that Edward Pease's home was at one time called The Ellens, in a remarkable provision embedded in the Second S&DR Act of 1824. It seems that the building of Skerne Bridge had a damming effect on the river to the detriment of the riverside gardens downstream. The Act provided that in respect of 'Land and Grounds called The Ellens belonging to Edward Pease of Darlington, the S&DR shall divert the River Skerne and make embankments to prevent The Ellens being flooded'. This is a classic example of Edward Pease's hard-headedness in business: despite his own investment in the S&DR he would not let that stand in the way of a claim against the company when its actions affected his interests elsewhere – in this case his precious gardens.

Joseph, Edward Pease's second son, was born at 146 Northgate on 22 June 1799. On his marriage to Emma, daughter of the Quaker Joseph Gurney of Norwich, on 20 March 1826,

---

JOHN PEASE 1797–1868

John Pease, the eldest child of Edward Pease, the first of five sons and three daughters, was born on 30 September 1797. Three of the brothers – Joseph, Henry and John – exercised a paramount influence in the affairs of the S&DR in its early years. 'They were, in effect, "managerial" directors responsible for a range of executive functions, and also for determining the company's strategic direction.' In its final years, John Pease led the S&DR delegation to the NER headquarters in York, which resulted in the amicable amalgamation that took place in 1863 (see below, page 00). John Pease married Sophia, daughter of Joseph Jowett of Leeds, on 26 November 1823, and the couple came to live in a large mansion called East Mount, so called because it was built on the east side of the River Skerne opposite his father's home. The gardens of both houses extended down to the river, where a footbridge gave access from the one to the other. In 1861, three years after his father's death, John asked the local authority to name the thoroughfare alongside his father's old home, 146 Northgate, Garden Street, a name it retains to this day. The street was well named because Edward Pease's extensive garden included orchards and vineries, in which he delighted in his declining years. A gate near the footbridge led into Paradise Walk along the banks of the river. The whole valley in this stretch was owned by members of the Pease family, so much so that it was nicknamed 'Pease-ful Valley'.

the couple went to live in a large house called Borrowses, which had been built by the Quaker Edward Backhouse in 1801. The mansion, which was renamed Southend by the Peases, was set in large grounds with an entrance in Grange Road. After many vicissitudes and changes of use it has become The Grange Hotel, replacing its namesake nearby on Victoria Road, which was demolished. It is located immediately to the south of the new stretch of road that links Victoria Road with Coniscliffe Road.

Edward Pease's brother Joseph was also born at The Grove, on 28 January 1772. When he married Elizabeth Beaumont on 1 July 1801 he bought the 8-acre Feetham's estate, to the south of the parish church, St Cuthbert's, and built himself a house, also called Feetham's, where he lived until his death on 16 March 1846. The son of Joseph and Elizabeth Pease, John Beaumont Pease, born on 27 July 1803, built North Lodge – opposite his Uncle Edward's house in Northgate – as his family home, where he lived with his wife, Sara Fossick, whom he married on 18 September 1825 – only days before the opening of the S&DR!

COMPANY OFFICES

In the early days, before the S&DR main line was completed, shareholders' meetings and sub-committee meetings were held on an *ad hoc* basis in rooms in the King's Head Inn. At the meeting that took place there on 25 May 1821, the committee authorised Joseph Pease to rent two rooms, at £5 per annum, on the upper floor of a shop, No. 9 High Row, Darlington, the premises of a cabinet-maker called James Marshall. These then were the first offices of the railway. These rooms were probably the ones referred to as the 'Railway-Office' in the handbill that announced the 'Order of Proceedings for the Opening of the Railway', and from which the bill (dated 19 September 1825), was issued, as the address is given as 'Atkinson's Office, High-Row, Darlington' at the foot.[8] The otherwise unknown name Atkinson, however, and the absence of a number, may indicate other premises nearby. This was a truly modest beginning for what was to become the administration centre of a highly complex enterprise.

In 1829 the offices were moved to elegant purpose-built premises in the classical style on the corner of Northgate and Union Street.[9] With the expansion of the company these offices were, in turn, extended, part of the first floor being given over to the directors' boardrooms. The building continued in use after the amalgamation of the S&DR with the North Eastern Railway (the NER) in 1863, until 1912. Although much obscured by later shop-fronts – in 1925 the ground floor was occupied by a 'Grand Clothing Hall' – and much altered through subsequent rebuilding, parts of the premises are still visible, the remainder having been demolished. Throughout its existence the administration of the S&DR was located in Darlington.

TICKETS, FARES AND PARCELS

The very first tickets issued to members of the public to travel on a steam-hauled railway were the 300 complimentary tickets issued to the 'gentlemen and strangers' for the inaugural train. The press notice published on 14 September 1825 invited applications for the tickets 'by addressing a note to the [company's] office in Darlington'. The notice is silent as to how the applicants were to receive their tickets, and time was short – only thirteen days remained

The Fighting Cocks Inn, near Darlington, from which tickets were sold in the early days of the S&DR.

to the date of the opening. It may be that a substantial number had already been sent to the invited 'nobility and gentry', and that those which were applied for, the remainder, were posted or, by implication, collected at the company's office. If so, in the strictest sense, No. 9 High Row was the world's first railway ticket office.

The first record of *fare-paying* passengers is contained in the timetable issued shortly after 10 October 1825 (when the passenger service commenced) for the company's only coach, *Experiment*: 'Mr Richard Pickersgill at his Office in Commercial Street, Darlington, and Mr Tully of Stockton will for the present receive any parcels and *book passengers*.' From the wording it appears that Messrs Pickersgill and Tully, while they were agents for the company, also carried on a business on their own account from their own premises in Darlington and Stockton respectively. It is almost certain that 'booking passengers' involved the issuing of tickets, so that for this first phase of the S&DR's passenger service tickets were issued from 'travel agents' in the two towns.

Within a few months, the operation of the passenger service was given over to private contractors. During this second phase, tickets to travel were issued either at the inns belonging to the proprietors, or at the pick-up points, which were also inns. (The poster announcing the commencement of the running of *The Union* coach listed no less than six inns.) In addition, a small number of inns and hostelries found themselves fortuitously situated by the side of the line as laid out, where passengers were encouraged to wait for the trains and where tickets were issued. The coaching inns at Fighting Cocks and at Urlay Nook are typical examples.

Railway stations and booking offices as we know them today were then non-existent. A plaque on a brick building near St John's Crossing, Stockton (now No. 48 Bridge Street),

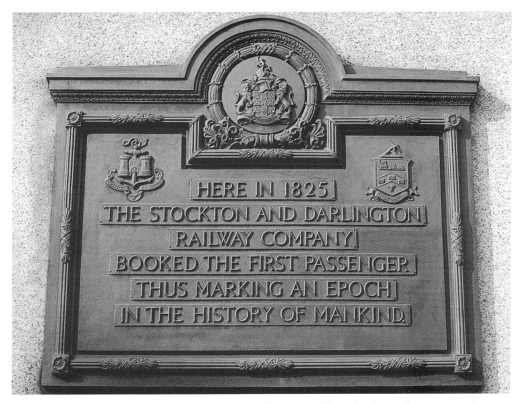

Commemorative plaque on the preserved S&DR booking office, 48 Bridge Street, Stockton-on-Tees.

unveiled in 1925, claims that: 'Here in 1825 the Stockton and Darlington Railway Company Booked the First Passengers. This Marked an Epoch in the History of Mankind'. This claim is now acknowledged as suspect. The building was in existence long before George Stephenson's *Locomotion* headed the first train down to the Cottage Row terminus on the quayside on that inaugural day. Tickets were certainly issued from this small office in later years, but these were probably coal-yard receipts. The inn that still exists at Fighting Cocks could claim to be the world's first railway ticket office – in the sense that it is the oldest building still standing by the lineside from which tickets were issued to passengers on the S&DR (or any other railway).

For some years after the introduction of locomotive-hauled trains in 1833 passengers sometimes travelled without tickets, paying their fares to the engine driver or guard. Shortly after his retirement, Jim Gowland, driver of the first passenger trains between Stockton and Middlesbrough, was interviewed by the *Newcastle Chronicle* and confided:

> There is not more than ten men in this town [Middlesbrough] who know where we landed our passengers in 1834. We landed them close on to the old steam packet wharf at Stockton Street, on a siding almost close on to the river. There was no station then, nor anything of the sort. No tickets were issued to the passengers, the guard took the money when he was there; and when he was not there I came off my engine and took it.

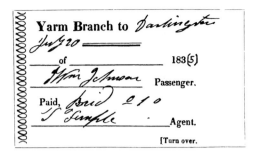

An early S&DR paper ticket, issued in 1835.

The first recorded passenger fares are those announced in the timetable for the company's coach *Experiment* – a flat rate fare of 1*s* for the 11 miles between Darlington and Stockton, or vice versa. This was without distinction of class, outside passengers paying the same as inside passengers. However, the handbill heralding the inauguration of the 'privatised' service provided by *The Union* coach, on 16 October 1826, announced a fare of 1½*d* inside and 1*d* outside per mile, that is 1*s* 4½*d* for the whole journey inside and 11*d* outside. With the introduction of the locomotive-hauled service in 1833 the two classes of fares were 2*s* and 1*s* 6*d* respectively for the whole journey, but fares on the slow or merchandise trains remained the same.

We know very little about the paper tickets that were issued in exchange for the fare, and which imitated the paper coupons issued by the stagecoach companies. They were printed with counterfoils in pads of one thousand. Issuing a ticket was a slow process, the clerk having to number the ticket and write in, on both ticket and counterfoil, the date of issue, time of departure and in some cases the fare and destination. One early example survives. It was issued to William Johnson for a journey from the Yarm branch to Darlington on 20 July 1835 at a fare of 1*s*. All this is written on one side, and on the reverse a standard printed regulation: 'This ticket to be delivered to the Engine-Man, expires with the date, and is not transferable.'[10] From that it would appear that in the early days, after acting as fare collector, the engine driver also did duty as ticket collector. When Joseph Pease was examined by the Commons Committee on Railways in 1839, he stated that the S&DR company's practice was to collect tickets at the start of the journey. 'Did that not encourage fare-dodgers?', enquired the committee:

> Do you find passengers take the short ticket, and travel all the distance?
>
> (Joseph Pease): We have found that but on one or two occasions: to remedy it, we have got them into a separate coach, drawing the bolt between the coach and the engine; and as they all agreed they were going half way, we left them to walk home, and that has cured it.[11]

No doubt that was a very salutary and effective method of preventing fare-dodging in future!

It is significant that from the beginning, as announced in these two posters, a parcels service is given equal weight with a passenger service. In the case of *Experiment*, 'Parcels are received and passengers booked' at the same place. In the case of *The Union*, 'passengers and parcels are booked' at any of the six named inns. In each instance, the company, or the coach proprietors, were not to be held accountable for any parcel more than £5 in value unless entered or paid for as such. This seems to refer to 'excess baggage' over and above the packages (in the case of *Experiment*) not exceeding 14 lb in weight, which were carried free. 'All above that weight to pay at the rate of 2*d* per Stone extra. Carriage of small parcels 3*d* each.'[12] Parcels 'requiring haste and care' were to be handed to Sarah Pearson, Glass Warehouse, Darlington; John Simpson, Majestic Office, Stockton; John Unthank, Middlesbrough; John Proud, Bishop Auckland; John Coxon, St Helen's Auckland; and Richard Thompson, New Shildon. From such modest beginnings developed the ubiquitous goods and parcels service that ultimately became a feature of every station in the kingdom.

It is clear that the operators were obliged to respond to the demands of this novel mode of transport, either in an experimental and *ad hoc* manner, or by adapting stagecoach practices to the new situation. It is interesting to note, however, that many of the customs and practices relating to passengers and parcels which were adopted in the first months and years of the S&DR have persisted through well-nigh two centuries up to the present day. The timetable for the company's coach *Experiment* announced that it would set off from the DEPOT at each place – that is, Darlington and Stockton – from 10 October 1825. At such an early date in the history of the S&DR it is most unlikely that these were stations in any sense of the word. On the contrary, it is probable that they were not buildings of any kind but merely 'depots' in the sense in which the word was used for the terminus of the coal branches on the opening day. If so, then the use of this word is another indication of the subordination of passenger services to freight services in the mind of the proprietors.

CARRIAGES

The first railway passenger vehicles were either existing stagecoaches with their road-running wheels replaced by smaller flanged rail-running wheels, or stagecoach style with small wheels constructed *de novo*. The fact that in both cases the work was undertaken by traditional coach builders[13] goes a long way to explaining why, to this day, railway vehicles have been known as 'carriages' and 'coaches' running on a 'railroad'. In respect of terminology at least, it was a seamless transition from one mode of transport to the other. (For a brief period in the early days the company was actually styled 'The Darlington and Stockton Rail*road*', as in the inscription on the bell of *Locomotion*.) Although the coming of the railways was to sound the death-knell for the stagecoach trade, the proprietors did not at first fully appreciate the extent of the revolution in passenger transport which their enterprises represented. This is the reason why many of the terms, the customs, and the practices of the horse-drawn coaching industry were from the first appropriated by the railways without question. Continuity was the order of the day, and not a break with tradition. For example, 'at any bends of the road, or other places where the view is obstructed, the coachman *blows a horn* to give warning of his approach to any wagons or vehicles which might be coming or going on the way'.[14]

These first railway coaches were designed to be hauled by horses as well as locomotives. (Passengers were always 'hauled' and coal 'led' on the S&DR.) Shafts were provided at both ends so that a single horse with its traces could be unhitched from one end and attached to the other thus enabling the coach to be drawn in either direction. We have a description of the first railway coaches, written by the Scottish clergyman who travelled the line in 1826:

> The coach consists merely of the body of a common inside and outside heavy coach set on a strong frame, with four wheels adapted for the rail-way, and considerably smaller than those of a carriage. The frame appeared too strong and heavy, and improvements might be made on this as well as on other parts of its construction, which seemed far from being the most suitable for this new mode of travelling. The coach had no springs of any kind, and yet the motion was fully as easy as in any coach upon the roads. . . . The coach never turns on the rail-way, but can be drawn either backwards or forwards with equal facility; the horse being merely unyoked from one side and yoked to the other, which is done in less than half a minute. To suit this arrangement, the front and back of the coach are made exactly alike, with the seats for the coachman, guard, and passengers, the same at either end, and the yoking place for the horse.[15]

The typical railway coach was 16 ft long overall including shafts, with a 4 ft 6 in wheelbase. According to the eye-witness quoted above, the coachman controlled an innovation not seen on stagecoaches – the brake:

> Such is the extreme mobility of the whole vehicle, and its load upon the railway, that when one is set in motion it is not easy stopping it; it is not enough here to 'pull up', according to the coachman's phrase; it requires an apparatus for the purpose, termed a brake, the operation of which is peculiar. It consists of a long arm or lever, turning on a centre between the fore and hind wheels at one side, reaching from thence up to the coachman's box, and having a short arm below, which, by moving the long one, can be made to press strongly on the rim of the wheels, and this creating a considerable friction, soon brings the carriage to rest. When the carriage is in motion, the long arm of the brake rests on a hook under the coachman's seat; and when he wishes to check the motion of the vehicle, or to stop it altogether, the driver unlocks the brake, sets his foot on the extremity of the long arm, and pressing the short one against the wheels, this instantly checks the motion, and gives him the complete command of the coach and the horse, lest these be moving ever so rapidly.[16]

When the S&DR took back into its own hands the operation of the passenger services in 1833, circumstances were rather different from the early years. Now the intention was to employ a new breed of tried and tested locomotives, displacing horses as motive power, and hauling trains of coaches coupled together. The existing rail-adapted coaches were unsuitable for this kind of operation in a number of respects. In particular, the permanent shafts at either end for harnessing the horses inhibited coupling-up. A new design of coach was therefore introduced which incorporated buffer-beams, couplings and other improvements. To mark this radical new departure in the design of passenger vehicles we shall here call them 'carriages' rather than 'coaches'. To inaugurate the service, a first carriage to the new design was commissioned and named *Union* (not to be confused with the privately owned horse-drawn coach *The Union* introduced in 1826). The new carriage commenced operations between Darlington and Stockton in November 1833. Although at first carriages were turned out in a variety of colours – blue, green and yellow – in February 1842 a standard livery of dark-red 'crimson lake' was adopted.

Unfortunately, none of these first carriages of the S&DR for the years 1833–45 have survived. Although they were rectangular in profile, the curved lines of the familiar stagecoach still appeared on the sides, betraying their pedigree – old practices die hard! These were composite carriages with a first-class compartment in the centre, flanked by a second-class compartment on either side. Each had two rows of inward-facing seats in the traditional manner. The first-class compartment was far better furnished and decorated than the second-class, non-upholstered compartments. They were provided with cloth linings, stuffed backs and seats, wooden armrests and plate-glass windows. However, the second-class compartments were infinitely better than the accommodation provided for the 'guard', who rode on a seat on the roof. (This does not necessarily mean that each carriage was individually hauled, or that every carriage in a 'train' carried a brakeman.) Luggage was also carried on the roof.

TIMETABLES AND PRECEDENTS

S&DR timetables are enlightening about the practices of the company, and its services for passengers, in its early years, as well as providing the times of trains. No doubt much of the detail was common to other railways of the period, but especially in the case of the first timetables we can discern the S&DR setting precedents that justify the title 'First in the World'.

1. 'No smoking in any of the Company's coaches' was the rule at least as early as 1840. It was not until 1868 that provision was made for smokers on British Railways. The situation has come full circle with the banning of smoking by most of the privatised companies at the time of writing.

2. A fare structure based on what the passenger could afford, or was willing to pay, for different standards of accommodation, was in place almost from the beginning. To some extent, therefore, the three 'classes', first, second and third, reflected *social* class. First and second class were derived from the 'inside' and 'outside' accommodation provided by the prototype stagecoaches. Whereas the lot of second-class passengers was improved, since they also were shielded from the elements in railway coaches, third-class vehicles for the poorer classes were at first little better than the open chauldron wagons adapted for the inaugural run. A vignette at the head of the winter timetable for 1837/8 illustrates these distinctions by depicting a train composed of two coaches of different styles and an open wagon bringing up the rear. The same timetable underlined class distinctions still further by providing for separate second-class and first-class trains between Darlington and Stockton. A rule-of-thumb fare structure for children had developed by 1847 in which 'children under three travel free; above three and under ten, if in First Class pay Second Class, if in Second, Third Class; no reduction in a Third Class carriage'. The regulations were not far removed from the present arrangement under which children below a certain age travel free, and children of school age travel at half the adult fare.

3. Special fares and special trains, but with conditions attached, were introduced from the 1840s. Thus 'Day Tickets, not transferable' at reduced fares (certain times) were offered in the 1846 timetable, which also advertises a fortnightly 'Market Train', which left Crook and Barnard Castle for Darlington at the early hours of 6.00 a.m. and 6.40 a.m. respectively. Times for so-called 'Merchandise Trains' were also given. These were 'mixed' trains composed of open wagons for goods consigned by customers, including livestock – horses and cattle – and 'flat' trucks for conveying customers' own carriages and gigs, bodily lifted

on and off. Thus the summer timetable for 1836, which commenced on 15 March, is headed 'S&DR Coaches: Increased Accommodation', and continues: 'Separate engines having been appointed for the conveyance of Passengers and Merchandise, and a coach attached the latter train, the Opportunities for Communication between the towns of Darlington and Stockton are doubled, and between Darlington and Shildon they are now four times a day.'

Three of the six daily trains each way between Darlington and Stockton were merchandise trains, for which second-class fares were charged. Horses, cattle and carriages were 'carefully conveyed' between Darlington and Stockton only by the merchandise train, the rates being $2s$ for a horse, $2s$ for a gig, $3s$ for a horse and gig, and $5s$ for a four-wheeled carriage. Whereas 'tickets must be taken at least 5 minutes before the train starts', horses and carriages were required to be at the station 'not later than 20 minutes before the departure of the train, and to prevent disappointment, one day's notice'. General merchandise 'for all parts of the Kingdom' was to be handed in to Robson Robinson of Commercial Street, Darlington. In the 1837/8 timetable the address was changed to the Merchandise Station, Darlington – the precursor of the goods depot. In addition to 'Special Trains' that appeared in the seasonal timetables, others were provided on an *ad hoc* basis to serve particular events, such as fairs and cattle shows.

4. The modern vogue for an 'integrated transport system' is nothing new. Early in its history the S&DR provided onward transport connections from its railheads, or stations en route, to tourist destinations not yet reached by the tracks. The 1836 summer timetable for the original main line advertised 'a car [*sic*] from Bishop Auckland to St Helen's Auckland and New Shildon, which meets each of the trains in going and returning'. Two years later, the timetable commencing on 20 March 1838 added an omnibus, which left Middlesbrough for Ormesby at 10.00 a.m., and a coach to Whitby and Loftus at 1.00 p.m. The timetable for 1837/8 provided connecting times for a coach called *Exmouth*, which plied between Stockton and Darlington, and Lancaster. A number of sulphur water springs were discovered in the neighbourhood of Darlington in the early nineteenth century, two of which led to the development of fashionable spas at Croft and at Dinsdale, the latter just upstream of Middleton St George on the River Tees. The nearest station to Dinsdale was at Fighting Cocks a mile to the north. The 'society' pretensions of Dinsdale Spa are apparent in the 1838 timetable, which advertised 'an Omnibus from Fighting Cocks to Middleton and Dinsdale [which] meets the *First Class* trains'. (First-class passengers were also courted in the timetable commencing on 1 October 1841: 'an omnibus, for each of the Principal Inns at Darlington and Stockton, attends the arrival and departure of the *First Class* trains'.) The 1838 timetable advertised coaches to Seaton, Redcar and Whitby (all coastal resorts) daily from Stockton, and from Darlington 'to all parts of the Kingdom'. The timetable commencing on 6 June 1846 covered an expanded network of lines as far afield as Barnard Castle and Crook. Thus a (road) coach is advertised leaving Barnard Castle for Bishop Auckland each morning, except Sundays, at 6.00 a.m. in connection with a train from Crook going east, and returning from Bishop Auckland on the arrival of the train going west to Crook.

5. During the railways' infancy, and before the institution of Greenwich Mean Time, considerable time-keeping difficulties were experienced from the custom of each local community observing its own 'sun time', which differed from place to place, east or west. (Stagecoach drivers got round the problem by using special watches that were set to run fast or slow depending on the direction in which they were travelling.) What was needed was the adoption of a standard time that would be the same for every locality served by the railway network. Indeed, it was the introduction of just such a 'Railway Time' in the 1830s, based on

the time as recorded in London, and which assumed that time was 'the same' all over the country, which paved the way for the *universal* adoption of Greenwich Mean Time. (It was fortunate that the growth of the railway network coincided with the development of the electric telegraph whose wires were first laid conveniently alongside the railway tracks. This made it much easier to synchronise clocks throughout the system.) The S&DR was among the first to realise the need for a standardised time, as evidenced by a footnote to the winter timetable of 1837/8, which commenced on 1 October: 'The Company's clocks are invariably kept with those of the GPO London, which time may be seen at Mr Harrison's, Clock Maker, High Row, Darlington.' Three years later, this facility was extended, for in the winter timetable of 1840/1, while Mr Harrison had been replaced by Mr Thompson, also of High Row, Darlington (near the company's office), a clock could also be consulted at Mr Buxton's, Market Place, Bishop Auckland, and in Stockton enquirers were directed to the Town Hall clock. These notices heralded the provision of clocks by the S&DR at its principal stations for the benefit of the travelling public and others – a considerable boon at a time when few ordinary citizens could afford the expense of a clock or watch.

6. Refreshment rooms were provided at North Road Station as early as 1845, according to the July timetable – another 'first'. Since the quality of railway refreshment room fare has since become the butt of innumerable jokes, one hopes that that tradition did not originate in Darlington!

7. The practice of allowing passengers to carry a limited amount of personal luggage free also originated with the S&DR. In 1846 the allowance was up to 100 lb – 'Merchandise not considered as such' – which was increased to 112 lb in the following year. The concession did not extend to accompanied dogs, owners of which were charged 1s regardless of distance – much as they are to this day. Parliamentary sanction for free passenger luggage was given in the S&DR (Consolidation) Act of 1849, with an increased allowance of 150 lb for first-class but remaining at 100 lb for second- and third-class travellers.

8. Finally, the modern practice of issuing separate winter and summer timetables to take account of seasonal differences in numbers travelling also originated with the S&DR. A timetable headed 'S&DR Coaches' makes the distinction: 'The Summer arrangements will cease on the 30th inst. and the trains will run the same as last season until further notice. Winter arrangements commencing 1st October 1840'. The practice had originated earlier, for a *winter* timetable for the 1837/8 season has survived for the main line, which by that time extended to Middlesbrough. No 'through services' were provided, passengers having to change – in winter! – at both Darlington and Stockton. Times were given for the three sections St Helen's Auckland to Darlington, Darlington to Stockton, and Stockton to Middlesbrough, and return.

PARLIAMENTARY TRAINS

The treatment of third-class passengers, especially their relegation to inferior rolling stock open to the elements, provoked a response typical of early Victorian reforming zeal. In 1844 Gladstone introduced his Regulation of Railways Act, which required all passenger railway companies to run at least one train daily each way on all lines, calling at all stations, at a fare not exceeding 1d a mile, the minimum overall speed to be 12 mph, and the accommodation to be protected from the weather. Receipts from such trains were not liable to passenger duty – a central government tax on all passengers carried by rail. It was imposed from 1832 and collected by the British railway companies for the Inland Revenue. Ostensibly, it was introduced to compensate for the loss of revenue caused by the railway companies' increasing inroads into road-coach traffic, which had been taxed for many years.[17] As Jack Simmons has pointed out, stagecoaches had not provided for third-class passengers, which was one of the railway's great innovations. With the passing of the Regulation of Railways Act, the minimum rights which it conferred on passengers put them far ahead of those who had travelled most cheaply by coach – for example the 12 mph limit was faster than the fastest coach, and the fare of 1d per mile was less than any coach had charged.

The railway companies varied considerably in their treatment of Parliamentary and third-class passengers. It is noteworthy that although the S&DR was obliged to abide by the Act, it was not obliged to include Parliamentary trains in its timetables, and did not do so. The public's patronage of these trains would depend on information passed by word of mouth. Nevertheless, they became a popular institution, despite the fact that by our standards this kind of travel was far from comfortable, and neither was it cheap – the penny was still a valuable coin to working men. In time, Parliamentary trains and third class together came to contribute an increasing proportion of the railways' revenue. In 1850 they accounted for nearly 30 per cent of the total; by 1860 the receipts from third class exceeded those from either first or second.

NORTH ROAD STATION

The first *permanent* steam-worked passenger service was introduced on the S&DR in September 1833. In anticipation of the event, the proprietors decided on 23 August to convert an existing goods warehouse in Darlington, built in 1827, into 'a place convenient for passengers waiting to take the railway coach' – that is, a booking-office cum waiting-room. So the S&DR's first station, and the world's first railway station in the proper sense of the term, was born. In November 1833 part of the building was converted into a shop and dwelling house. They were let to May Simpson at £5 per annum for a trial period of one year on condition that 'she kept the coach office clean and afforded every necessary accommodation to coach passengers'.

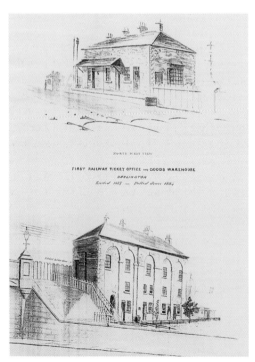

The first railway ticket office and goods warehouse alongside North Road, Darlington. Opened in 1827, it was demolished 1864.

The station was situated just to the east of the point where the North Road turnpike crossed the railway. In 1856 a bridge replaced the crossing, the roadway being scooped out to pass beneath it. This bridge lasted less than twenty years and was replaced by a stronger bridge, built in 1876 and still in existence today. The foundations of this first station, which was demolished in 1864, can be seen just to the east of the bridge on the south side of the track.

In 1842 the converted warehouse was replaced by a purpose-built station 350 yards

North Road Station, Darlington – opened in 1840 – as it appears today.

to the west of the North Road crossing. The fact that it has subsequently been modified and extended probably gave rise to the misconception – repeated in some histories – that it is the *second* station to be built on this site. (Two vertical arrows on the inside wall to the left and right of the main entrance mark the 'Extension of the Original 1842 Station, 1888'.) Throughout its history the station has been known as 'Darlington North Road'. The south elevation today presents a plain but attractive late Georgian facade with a two-storey centre (the original station) and one-storey wings (the extensions). The centre has an attached colonnade with cast-iron columns.

Internally, the station consisted of a train shed and a carriage shed, separated by a wall. At first there was only one through platform. To mark the 150th anniversary celebrations the former train shed of North Road Station was converted into the Darlington Railway Centre and Museum. It was opened on 27 September 1975 by HRH The Duke of Edinburgh, 150 years to the day after the inaugural run. It now houses a large collection of items illustrating the history of the S&DR. Of particular interest is a three-dimensional panorama of the original S&DR line from Brusselton summit to Stockton as it appeared in 1825. Some 70 ft long, it incorporates detailed scale models of the principal buildings and other features along the line – including North Road Station.

Shortly after North Road Station was opened in 1842 a companion brick goods depot was built nearby, to the south-east. This took the place of the original 1833 warehouse, and thankfully survives to this day. It consists of two parallel sheds with pitched roofs and a central square clock tower. It is presently used as a locomotive restoration workshop by the Darlington Railway Preservation Society. Other buildings in the vicinity of the station are dealt with in the next chapter.

CHAPTER 6
The Rise of the Railway Industry

THE COMPANY'S LOCOMOTIVE WORKS

The S&DR built its first locomotive works at New Shildon on a site alongside the line, strategically situated at a point where static steam haulage gave way to locomotive haulage at the foot of the Brusselton east incline, just to the west of the road crossing at the Mason's Arms. Originally there was only one engine shed with space for two locomotives, together with a long narrow shed occupied by the joiners and blacksmiths. The initial workforce of only twenty men was soon expanded to fifty. Strictly speaking this was not at first a locomotive *building* works, but a repair and maintenance depot. For some years, the company continued its initial practice of contracting out the building of locomotives for the line. The first four – *Locomotion*, *Hope*, *Black Diamond* and *Diligence* – were built by Robert Stephenson and Company at their Forth Street Works, Newcastle. Of the next 26 locomotives, 9 were built by Timothy Hackworth while employed by the company, and 17 by outside contractors including Robert Stephenson and Company (9), R. & W. Hawthorn (4), William Kitching (3), and William Lister (1).

The Mason's Arms, Shildon. The inaugural train hauled by Locomotion No. 1 *began its journey here on 27 September 1825.*

Timothy Hackworth, Locomotive Superintendent

It is an irony of history that in respect of its locomotives the S&DR owed a greater debt to Timothy Hackworth than to George Stephenson. The 'Father of Railways' had little to do with the S&DR after the construction of the line because of his commitments elsewhere. It is a further irony that these commissions were largely the result of the fame that George Stephenson acquired following the success of the S&DR, in which he had played a leading role. Timothy Hackworth did not enjoy the same degree of renown during his lifetime. Nevertheless, the developments initiated by this 'talented engineer endowed with immense practical abilities' led to the emergence of a type of goods engine on which the prosperity of the S&DR ultimately depended.

The lives of Hackworth and Stephenson had much in common. Both were born in Wylam – Stephenson on 9 June 1781, and Hackworth on 22 December 1786. Both worked as engineers at Tyneside collieries, and both were involved there in the development of the steam locomotive. (Hackworth was foreman-blacksmith at Wylam Colliery while William Hedley was experimenting and building his locomotives there.)

When the S&DR was opened on 27 September 1825, Thomas Storey, one of the two resident engineers under George Stephenson's direction, was promoted chief engineer. As it was to turn out, however, it was another early appointment that was of far greater consequence for the company's fortunes. On 28 June 1825, on the strong recommendation of George Stephenson and William Patter, his former employer at Walbottle Colliery, who described him as 'ingenious, honest and industrious',[1] the company appointed Timothy Hackworth its first locomotive foreman – and hence the world's first 'shedmaster' – four months before the delivery of *Locomotion*. There are other versions of his title, and those of his successors, including 'Superintendent of the Permanent and Locomotive Engines', 'Locomotive Superintendent', 'Resident Engineer', and 'Superintendent Engineer'. These titles reflected changes in the nature of the appointment over the years. For example, in November 1828, on the eve of a deputation to the S&DR from the board of the L&MR to evaluate the benefits of locomotive power as against fixed haulage, Edward Pease wrote to Timothy Hackworth:

> I am informed that a deputation is coming from Liverpool to see our way, but more particularly to make enquiry about Locomotive power. Have the engines and men as neat and clean as can, and be ready with thy calculations, not only showing the saving, but how much more work they do in a given time. Have no doubt wilt do thy best to have all sided and in order *in thy department*.

Timothy Hackworth's starting salary was £150 per annum, 'the company to find a house, and pay for his house, rent and fire'.[2] (While the company were finding a house, Hackworth took up temporary residence in Darlington before moving to New Shildon.) In addition to maintaining the locomotives and the static steam engines, he was also responsible for the construction and maintenance of the chauldron wagons and, in due course, the company's passenger coaches.

The working of the first four engines, of the 'Locomotion' class, was unreliable and far from satisfactory. One of Hackworth's first tasks, therefore, was to improve the working of the company's steam locomotives. Early in 1827 he was engaged in the design of no less than three new locomotive types, and it was in this context that the management committee authorised him to construct a locomotive that would 'exceed the efficiency of horses'.[3]

TIMOTHY HACKWORTH 1786–1850

The name of Timothy Hackworth will always be identified with Shildon in County Durham but, like his contemporaries George Stephenson and William Hedley, Hackworth's roots were on Tyneside. He was born at Wylam, where he became foreman-blacksmith at the colliery there, before moving on to Walbottle Colliery. He was probably employed at the Forth Street Works of Robert Stephenson and Company in Newcastle while Robert was absent abroad. On the strong recommendation of George Stephenson he was recruited as the S&DR's first locomotive foreman in June 1825, four months before the opening. He remained in Shildon for the rest of his life. Timothy Hackworth was 'a talented engineer . . . with immense practical abilities'. He designed and built *Sans Pareil*, which came a worthy second in performance to *Rocket* in the Rainhill Trials, and *Royal George*, which saved the day for locomotive traction on the S&DR. He devised practical solutions to many of the unprecedented problems posed by the first railways, including a double-acting winding gear for inclines, and the robust 'plug wheel' for steam engines. When Hackworth left the direct employment of the company in 1840 to set up as a general engineer and locomotive builder and repairer in his own right, he established the 'state of the art' Soho Engine Works at New Shildon. There he built model terraced housing for his employees as well as his own home, now converted into the Timothy Hackworth Victorian and Railway Museum. A staunch Methodist, Hackworth's concern for the spiritual and social welfare of the men and their families led him to provide a chapel and the world's first railway institute. Shildon is understandably proud of its favourite adopted son. Streets, institutions, and public houses bear his name. A life-size statue stands in the town centre and everywhere *Royal George* appears as an emblem and an unofficial coat of arms. He died on 7 July 1850 and lies buried in the graveyard of St John's, Shildon's parish church, where a tombstone records his many achievements.

Timothy Hackworth's home, Soho Works, Shildon, now occupied by the Timothy Hackworth Victorian and Railway Museum.

Locomotive No. 5, *Chittaprat*, had been bought from the builder Robert Wilson, whose works were near to those of Robert Stephenson and Company at Forth Street, Newcastle, at a bargain price of £380 because it seemed to have a good boiler. (It was the workmen of the S&DR who gave it its unusual name, because of the peculiar noise it made.) No official drawings exist of *Chittaprat*, but we do have a notebook sketch and details from Marc Seguin's visit to the railway in 1826. Timothy Hackworth made use only of the engine's boiler, which he enlarged, and possibly the wheels and some of the valve gear. The remainder of what became the celebrated and highly acclaimed No. 5 *Royal George* was entirely his own build and invention.

Royal George was the first locomotive to have three axles with the wheels coupled via outside coupling rods. These properties, together with its weight – 12 tons 7½ cwt compared to 8 tons 5 cwt for the first generation of Stephenson-type engines – gave it the marked advantage of greater tractive adhesion.

The controls were arranged so that the driver stood on a footplate conveniently behind the boiler instead of precariously beside it, as was formerly the case. However, because a return-flue boiler was used, the driver was even further removed from the fireman who worked from the chimney and firegrate end. Because the boiler feed-pump was under the driver's control, the water tender was obliged to be at his end of the engine with a separate coal tender at the other.

Royal George commenced work in November 1827 with George Gowland as its driver and 'at once proved a triumph for Hackworth'. It was truthfully described in a contemporary report as 'undoubtedly the most powerful [engine] that has yet been made'. So pleased was the company with the engine that it gave Hackworth a bonus of £20, and *Royal George* was soon hauling loads of up to twenty-three coal wagons. During its first fifteen days in traffic, it averaged 36 miles a day, hauling an average load of 46½ tons. It was undoubtedly a far larger and more powerful locomotive than any of its predecessors on the railway, with a boiler operating at 52 lb per square inch, and 141 square feet of heating surface compared with *Locomotion*'s 60.

From the beginning, Hackworth's *Royal George* proved more reliable and economic than George Stephenson's earlier locomotives.[4] It was the first of a new generation that would leave its mark on the development of the steam locomotive. Visually, the most obvious characteristic was that the locomotives had a tender at each end. *Royal George* was the precursor of a number of these highly successful heavy six-coupled goods engine types, typified by the 'Majestic' and 'Director' classes, which for decades were the mainstay of the S&DR. They were particularly suited to hauling trains of coal wagons, and were built for the railway from 1829 onwards by Hackworth as locomotive superintendent. The fortunes now began to turn distinctly in favour of the steam locomotive.[5]

It was Hackworth, too, who invented the so-called 'plug wheel', in order to deal with the problem of the perpetual breakages of the cast-iron wheels on both locomotives and rolling stock. Hackworth's wheel was cast in two parts, each of matching form: 1. an inner circular portion, which was accurately machined to run true on the axle, to which it was attached by keys; 2. an outer portion, which was secured to the centre by wedge bolts and eight to ten wooden taper plugs (hence the name) driven into holes between the two halves. This ensured that the periphery of the wheel could be accurately trued up with the centre. When the wheel had been assembled in this way, a wrought-iron tyre was shrunk onto it. A principal advantage of Hackworth's innovation was that either the outer or the inner half of the wheel could be separately removed, as they frequently needed to be in practice. In short they were strong, made for ease of replacement, and became standard on all S&DR engines. Hackworth's wheels were made by Michael Longridge at the Bedlington Ironworks, and in one of the letters between the two men there is a marginal sketch by Hackworth clearly revealing that his tyres – or 'hoops' as they were then called – were plain rings, like the metal tyres of a cartwheel, shrunk round the tread of the wheel outside the flange, the latter being part of the wheel casting.[6]

Timothy Hackworth's association with the S&DR was a peculiar one. In October 1829, while still in their direct employment, he was permitted to design and construct his own entry, the locomotive *Sans Pareil*, for the famous Rainhill Trials on the L&MR. This was in direct competition with George and Robert Stephenson's own entry, the victorious *Rocket*. Whether *Sans Pareil* was in fact the better engine has been the subject of much controversy, but its failure in the trials represented one of the few set-backs suffered by Hackworth in his long career. *Sans Pareil* itself proved to be a very satisfactory engine. It was sold to the Bolton and Leigh Railway, where it put in many years of excellent service.

From 1833 onwards, when the management committee decided to change the system, Hackworth left the company's direct employment and entered into a contract for the working of the railway, including the locomotives, the workshops, the tools and the machinery. Although the day-to-day operation of the engines was in the hands of the traffic manager, John Graham, Hackworth was responsible for providing and paying the drivers. Their experiences under his regime are graphically described in the following account by James Gowland, brother of William Gowland, the driver of both *Royal George* and *Sans Pareil*:

> We had a farthing per ton per mile. . . . For that figure we found everything except the engine, coal and oil and firemen. . . . An engine could then make perhaps 10*s* per day after all was paid. We were paid better than anyone else and we always had plenty of money. Some men took a contract with an engine. They paid the enginemen, say, 5*s* and the fireman 3*s* 6*d* per day, and then there was a grand living for doing nothing. . . . I was 27 times to Middlesbrough for Harry Joyce in one week. Harry had been having a drinking bout, and he never came near the place. He paid me 5*s* per trip, and I made a good week's wages, but then I was working night and day.[7]

This aspect of Timothy Hackworth's responsibilities ceased when, on 1 March 1837, the management committee changed the system yet again and took over the direct management of all haulage operations on the line.[8]

Timothy Hackworth continued to use the company's Shildon Works where he had, until then, been an employee. Now that he was no longer directly employed by the S&DR he was free to expand his private business interests, and to establish his own general engineering and locomotive building works, to supply engines under contract for both the S&DR and its commercial rivals. This was entirely in line with the S&DR's policy, which it maintained to the end of its independent existence, of ordering locomotives from contractors as well as building them in its own works. At this time, in addition to Hackworth, other principal contractors were William and Alfred Kitching, and William Lister, who both had premises in Darlington, and R. & W. Hawthorn of Newcastle upon Tyne.

THE SOHO WORKS

The site chosen by Timothy Hackworth for both his works and his new home was a strategic one, half a mile to the east of the S&DR's locomotive works, in the angle between the Surtees Railway and the Black Boy colliery branch, and therefore within a few yards of the main line. The complex commenced operations in 1840 when it was christened the 'Soho Works', it is believed at the suggestion of Joseph Pease. The Birmingham works of Boulton and Watt carried the same name from about 1760, as did the first static steam engine works in America before the turn of the century. For the first seven years, the works was largely run by Timothy Hackworth's brother Thomas, while he himself was engaged on other S&DR affairs. The registered name of the company was Hackworth and Downing, the Hackworth being Thomas, with Nicholas Downing his partner.

In May 1840 Timothy Hackworth ended his contract with the S&DR to concentrate on the running of his Soho Works, building locomotives and stationary steam engines for the S&DR and other railways, as well as winding engines for colliery companies, and hydraulic presses and grinding mills for local engineering concerns.[9] As Maurice Kirby has emphasised, when his own business took precedence, Hackworth had no hesitation in subcontracting work to others.[10] In this respect, he made particular use of W. & A. Kitching, who also engaged in repair and maintenance work for Hackworth. The first locomotive to be built there for the S&DR was 0-4-0 No. 41 *Dart*, delivered in April 1840. It was withdrawn at the end of 1879 in the days of the NER.

WILLIAM BOUCH: LOCOMOTIVE SUPERINTENDENT

When Timothy Hackworth's contract ended in 1840, the S&DR's 'New Shildon Works' were taken over by Oswald Gilkes and William Bouch (1813–76).[11] In about 1860, Bouch was promoted 'Locomotive Superintendent and Engineer of the Railway', at a salary of £450 per annum, a position he held with both the S&DR and then the Darlington Committee of the NER until his death in 1876. William Bouch was responsible for the design of one of the classes of goods locomotives for which the S&DR was noted. This was a long-boilered 0-6-0, which originated in 1854 when the Middlesbrough firm of Gilkes, Wilson and Company built the first two of the class. They had boilers with the firebox overhanging the short-coupled wheelbase, and represented the final examples of the Stephenson long-boilered design. (One of the engines, No. 1275, is preserved at the National Railway Museum at York.) Later, in 1860, Bouch was also responsible for the introduction by the S&DR of the first bogie 0-4-0 locomotives in this country, *Brougham* and *Lowther*, which were noted for their large cabs.

Significantly, William Bouch served his apprenticeship with Robert Stephenson and Company in Newcastle. With the approval of the proprietors of the S&DR he 'privatised' their original locomotive works near the Mason's Arms crossing, which was his base, by forming 'The Shildon Locomotive Company Limited'. The company continued under Bouch's direction, although it was in reality simply an S&DR subsidiary. It promised to fulfil all the commitments entered into by its predecessor.

Timothy Hackworth died on 7 July 1850 after a very sudden illness.[12] His son John Wesley Hackworth carried on the business for a while, but the works were soon abandoned. On 9 May 1853 they were put up for auction by Timothy Hackworth's heirs and bought for £4,900 in July 1853 by the S&DR itself. From that time, therefore, the company controlled two locomotive works in Shildon on separate sites. Although the administrative headquarters of the S&DR was in Darlington from its inception, the centre of gravity of its locomotive and engineering works was in Shildon, although this would also move to Darlington in due course.

The Soho Works continued in railway use until 1883. What remains of them, together with Timothy Hackworth's house and adjacent workmen's cottages, now constitute the Timothy

The pattern shop and storage shed, Soho Works, as it appears today.

Hackworth Victorian and Railway Museum, opened by the Queen Mother on 17 July 1975. Near the house the original pattern shop and storage shed now shelters a replica of *Locomotion* and early S&DR rolling stock. The paint shop section at the rear had underfloor central heating, now restored – an innovation way ahead of its time. There are no other remaining visible evidences of the once-bustling Soho complex. The two-storey erecting shop, with the pattern-makers above, and the combined foundry, machine shop and blacksmith's shop, have all disappeared.

THE DARLINGTON WORKS

Within a few years of the S&DR's acquisition of Timothy Hackworth's enterprise at New Shildon to augment its own works there, the two sites together proved to be too small for their purpose. In 1854 John Dixon, then locomotive superintendent and engineer, suggested that they should be re-sited. Three years later, a decision was taken to build at Darlington, in the name of the S&DR's subsidiary, the Shildon Works Company. The transfer of the S&DR's locomotive operations from Shildon to Darlington signalled a marked change in the related fortunes of the two towns. Although Shildon did not decline, Darlington was henceforth the star in the railway engineering firmament, and Shildon was its satellite.

The new works, totalling 6 acres on a 20 acre site – for the fabrication, erection, repair and maintenance of the company's locomotives – were opened on 1 January 1863. Ironically, a bare six months later, the S&DR was amalgamated with the NER. The new works soon became known affectionately as the 'North Road Shops', a label that persisted throughout their lifetime. Some 150 staff were transferred from Shildon, including the locomotive superintendent and engineer, William Bouch. The first locomotive to be out-shopped, in October 1864, was No. 175, named *Contractor* in honour of the Darlington Quaker David (later Sir David) Dale, who had done much contract work for the railway. Some time later, Dale and Bouch went into partnership. Some locomotives continued to be out-shopped from the Shildon Works for a time, but they were merely assembled from parts made at and supplied by the North Road Works.

North Road Shops eventually expanded to become Darlington's largest employer and one of the largest locomotive-building works in the kingdom, if not worldwide. In its heyday as part of the L&NER empire, the works employed 3,500 men. Sadly, it closed completely on 2 April 1966, with the loss of 2,540 jobs, after a century and more of continuous activity, the irrevocable end of an era in Darlington's railway engineering history.[13] Much of the site was redeveloped as a Morrison's supermarket.[14] The works' magnificent clock, by which generations of hurrying employees timed their arrival and departure, has thoughtfully been retained and restored. It overarches the North Road pavement, attached to the outside wall of what was once the electrical shop and drawing office. It appears to be suspended in mid-air, at the end of an elaborate, floreated cast-iron bracket. As time passes, it becomes increasingly difficult to appreciate that this site was once at the heart of Darlington as a railway town.

The first S&DR engine shed to be built in Darlington was located at Whessoe Road, immediately to the north of the present Hopetown Junction. In 1877 it became a paint shop and then, in NER days, a workshop for electric locomotives. The shed still stands and was last used as a local signals and telecommunications store.

In 1853 the S&DR terminated its contracts with private concerns and built workshops for the construction, maintenance and repair of its carriages on a site immediately to the south-west of North Road Station, alongside and to the west of the Coal Depot branch near to its junction with the main line. The premises backed onto Hopetown Lane. They were eventually inherited by the L&NER and then for many years lay derelict until they were acquired by the A1 Steam Locomotive Trust in 1996. They have been restored and converted into a locomotive works, initially for the erection, operation and maintenance of the Trust's newly built steam engine, an A1 'Peppercorn' class Pacific, No. 60145 *Tornado*.

ANCILLARY INDUSTRIES

Private locomotive builders were already established in Darlington before the S&DR began to build the North Road Shops there in 1863. Since the company's headquarters was in the town, and given its policy of contracting out the building of new locomotives, it is not surprising that private enterprises established themselves in Darlington, the railway's administrative and geographical centre, and adjacent to its main line.

In 1790 William Kitching, a Darlington Quaker, opened an ironmonger's shop in Tubwell Row, with a small foundry behind. In 1824 the supplying of nails to the S&DR marked the start of the firm's long involvement with the railway industry. In 1830, under the name William and Alfred Kitching, it moved to a site at Hopetown on the south side of the S&DR line just to the west of the Darlington Coal Depot branch junction. Here the foundry was soon supplying a first batch of locomotives under contract for the S&DR – No. 25 *Enterprise* in 1835, No. 29 *Queen* in 1837, and No. 30 *Raby Castle* in 1839. Although the final locomotive supplied to the company just before its demise in 1864 was numbered 175, the actual total of S&DR locomotives was many more because of the practice of reallocating numbers when a

Locomotive No. 25 Derwent, *built by A. & W. Kitching and Company for the S&DR at their Darlington Works in 1845, as preserved at the Darlington Railway Museum.*

former holder was sold or taken out of service. One such locomotive was No. 25, *Derwent*, manufactured by Kitchings in 1845 and which inherited its number from the locomotive *Enterprise* when that was withdrawn. Happily, *Derwent* has been preserved as the concluding form of the very many variations in design that Timothy Hackworth's 0–6–0 'Royal George' class and its successors underwent during their long and distinguished career. Altogether W. & A. Kitching supplied sixteen locomotives for the S&DR, as well as undertaking much maintenance and repair work. The supply of locomotives to other railways, and diversification into other branches of engineering, led to a considerable expansion on the original site.

The history of W. & A. Kitching highlights many of the factors that favoured Darlington's transformation into a railway town. At the dawn of the railway era, a number of small forges and similar light industries existed in Darlington to cater for the needs of not only a large agricultural hinterland but also the woollen and linen industry, which was well established here. There was thus an existing reservoir of light-industrial skills, tools, premises and processes that could readily be called upon to nurture the railway industry in its infancy.

Incidentally, with many of the major industries with which the S&DR was associated – coal mining, woollen mills, locomotive building, and the operation of the railway itself – there was a good deal of commercial cross-fertilisation and mutual self-interest. We are not surprised to discover, therefore, that William Kitching was a shareholder in, and a director of, the S&DR, for which his firm supplied locomotives.

Another locomotive building works established nearby, even earlier, in 1830, belonged to the firm of William Lister. In that year, Listers was responsible for the rebuilding of the

S&DR's third engine, *Black Diamond*, which was originally supplied by Robert Stephenson and Company of Newcastle to the design of *Locomotion* in 1826. The firm subsequently supplied three locomotives for the S&DR.

In Chapter 9 we will trace the fortunes of the rival companies that invaded the territory of the S&DR during its lifetime. Two of them together established a south–north route that traversed the west–east main line of the S&DR in an area known as Albert Hill. This conjunction of routes had two important consequences. First, a number of forges and the like were established in the mid-nineteenth century in the Albert Hill area. Their prime (although not their only) purpose was to serve the light engineering needs of the rival railway companies. It was mainly for this reason that they were located at Albert Hill – the only place that had ready access to both of their principal customers.

The second, and much more important, consequence was the incentive for railway-related industries, and for heavy engineering companies in general, to locate alongside the two rail-transport arteries. So, in the course of time, two 'corridors' of industry developed in Darlington, at right angles to one another, centred on Albert Hill. As the epitome of a railway town, Darlington was shaped by this conjunction of routes and the industries that became their neighbours.

CHAPTER 7

Main-line Branches and Extensions

Branches

Before we consider in detail the branches from its main line which the S&DR was authorised to build, it is important to note that the future development of the company was circumscribed in one important respect by a particular provision of the First S&DR Act of 1821. The owners of land within *5 miles* of the main line, either individually or in association with other landowners, were given the right to lay down their own collateral branches to communicate with the S&DR's line. The company was obliged to allow these alien branches to effect a junction with its line provided that the owners bore the cost of the necessary points/switches. Specifically, the S&DR 'shall not receive any Tonnage for the passage of any goods etc. along such Branches'. This provision was evidently designed to safeguard the interests of the proprietors of coal mines in the neighbourhood of the railway, but it set a precedent. In effect, a corridor 10 miles wide was established within which landowners and mine proprietors could exploit, with a minimal outlay on track, a subsidised transport system brought to their doorstep by the S&DR. (It has to be said that many of them would be S&DR directors or shareholders themselves). Another consequence was the establishment of two categories of branch line which differed in status. The premier branch lines were those of the railway company itself, with routes specified in its Enabling Acts. The numerous 'private' colliery branch lines were those that required no authorisation other than the blanket provision quoted above.

The outcome of successive Parliamentary Proceedings, as they related to authorised branch lines, may be summarised as follows:

FIRST S&DR ACT: 19 APRIL 1821
1. Darlington Coal Depot branch
2. Yarm Coal Depot branch
3. Coundon Turnpike Gate branch   Not proceeded with
4. Norlees House – Evenwood Lane branch   Not proceeded with
5. Stockton branch   Not proceeded with

SECOND S&DR ACT: 23 MAY 1823
6. Black Boy branch   Substituted for No. 2
7. Croft Depot branch

THIRD S&DR ACT: 17 MAY 1824
8. Haggerleases branch   Substituted for relinquished No. 3

Let us now consider the subsequent history of the five branches that were proceeded with after their authorisation by Parliament. Three were intended to serve coal depots, urban and rural, and two were intended 'to effect a facility of communication between certain collieries and the main line'. Although the Black Boy, Croft, and Haggerleases branches were all authorised by 17 May 1824, their building was suspended because of the parlous financial state of the S&DR for some years after the opening.[1]

DARLINGTON COAL DEPOT BRANCH

George Stephenson brought his main line much closer to the northern outskirts of 'the Township of Darlington' than Overton's route. Although, in view of its original purpose, this seemed to render the Darlington branch superfluous, the opportunity was taken under the head of 'Branch Line Deviations' to substitute what became known as the Darlington Coal Depot branch. It was newly defined in the 1823 Act as 'commencing at George Allen's field [and proceeding] to or near to certain Streets in Darlington called Union Street and Bondgate'. The intended branch therefore forked just short of the town centre to reach two termini. In the event, Stephenson chose to shorten the branch as authorised: according to plan, it diverged from the main line at a point immediately to the west of the (later) North Road Station, but after 300 yards it was diverted to run alongside Hope Town Lane and terminated near to Bondgate Bridge, precisely where Overton's proposed branch was shown to terminate in Overton (2). That intended branch was 1,540 yards long, Stephenson's published branch 1,230 yards, and the branch as built 800 yards. It was the only one of the five branches authorised in the First S&DR Act of 19 April 1821 to be opened on the same day as the main line, 27 September 1825.[2]

YARM COAL DEPOT BRANCH

The second Coal Depot branch diverged at a spot, designated the original Yarm Station, at the level-crossing on Elton Lane, near what is now Allen's West Station. It was 1.2 miles long and headed south and then east to terminate behind the New Inn, later renamed the Railway Inn and more recently the Cleveland Bay Hotel beside the A135 road close by its junction with the A67 to Goosepool and Darlington.[3] On the right-hand side of the road (travelling towards Darlington) the coal depot manager's house still stands, identified by a standard S&DR property plate inscribed D 13. The stone wall alongside the road here is capped by the later type of S&DR stone sleeper blocks. Both the branch and the New Inn were opened on 17 October 1825. Although the coal depot was sited on the Egglescliffe village side of the River Tees, that is in County Durham, it was intended to serve the inhabitants of Yarm on the Yorkshire side. We recall that some of Yarm's most prominent citizens had been instrumental in promoting the railway, as recorded on a plaque on the Town Hall in the High Street.

The first coals for Yarm were carried down the branch on 11 October 1825, soon after the opening of the main line. On 17 October the depot was opened to the public, followed by a celebration dinner that same evening at the New Inn, when a toast was proposed to 'the Gentlemen of Yarm'. Passenger traffic commenced on 16 October 1826.

From the opening of the branch in 1825 until 1833 it seems that the main-line horse-drawn passenger service was diverted to travel down the branch and back to the junction station before resuming its journey – thus adding considerably to the overall time. When steam locomotives replaced horse power in 1833 it was decided that they would call only at the junction station and not proceed up and down the branch, enabling the timetable to be speeded up. The junction station itself was closed on 16 June 1862. In the following year, it was decided to transfer the waiting shed and the platform to Aycliffe Lane (later Heighington), although the clock was to remain as it was 'likely to be useful to the Mineral and Goods enginemen'. The branch was finally abandoned in 1871.

BLACK BOY BRANCH

This line was initially built to serve two collieries of that name (themselves named after a nearby public house), 'in the township of Coundon'. Rather confusingly, the first, which operated from 1810 to the 1830s, was called Old Black Boy or Coundon Gate and the second, which operated from 1825 to the 1860s, was called Black Boy or Coundon Grange, being renamed Old Black Boy when the first colliery of that name ceased to operate. This 2⅓ mile branch began at Shildon Junction, half a mile east of the Mason's Arms crossing, where the line passed between the surviving stables, bank-riders' cabin, and coal drops served by a siding from the branch.[4] This was an interchange where rope haulage of the loaded and empty chauldron wagons gave way to horse haulage (hence the stables), and later locomotive haulage, on the main line. The branch ended at Coundon Gate Colliery,[5] which, despite its name, was situated half a mile to the south of Coundon Turnpike Gate, the terminus of the branch as authorised in the Act of 23 May 1823. In fact, that part of the route was never completed.[6]

Since it was required to surmount the high ridge to the north of New Shildon, the branch was powered by two stationary steam engines at the summit at High Shildon (similar to those at Etherley and Brusselton), which operated an ascending incline with a gradient of 1 in 20 from Dene (or Denburn) Beck and a descending incline also with a gradient of 1 in 20 to Shildon Junction.[7] The valley slope from the terminus at Coundon Gate Colliery to Dene Beck was worked down by gravity and up by horse power. The ascending and descending inclined planes were also initially worked by horses since the summit winding engines were not then completed. Although they commenced operations in 1828, they proved inadequate and were replaced in 1835 by new engines of 50 hp built and installed by Timothy Hackworth. For some years, the S&DR found it economic to contract out the working of the Black Boy and Brusselton inclines at the rate of .47d and .59d per ton-mile respectively.[8]

The opening of the line stimulated the development of further pits – some of which were named 'Black Boy' by association, to add to the confusion. Thus the Black Boy Colliery (Gurney Pit) was served by the 'Black Boy Private Colliery Railway', which diverged from the main Black Boy branch shortly after crossing the Dene Beck, and executed a sharp right-angled bend to run east up the valley, on a self-acting incline powered by a stationary engine at the head.[9] Close by the junction of this branch with the Black Boy line proper an existing colliery, opened in 1825, took on the name Machine Pit – Old Black Boy.

Another sub-branch diverged from the summit level at Chapel Row, where a coal depot was established, and ran north-east to the South Durham – later Old South Durham – Colliery (1830–60).[10] It commenced operations in 1833 and descended to the Dene Beck by way of an incline powered by a stationary steam engine situated at 'Eldon Bank Top'. The sub-branch was joined at this point by another from 'Eldon Pits', opened out in the 1820s, further to the north-east. Halfway down the Black Boy northern incline another sub-branch diverged to the west which served the Deanery Colliery (c. 1810–40) and a little further on the Adelaide's 'Shildon Bank Colliery' (1830–1924), owned at the time by Joseph Pease.

In 1826, because of the delay in completing the branch, the company felt obliged to compensate the owners of the Black Boy Colliery at the rate of 1^s 6^d per ton for all coal delivered to the railway terminus by road. A little later, compensation was extended to William Lloyd Wharton, the lessee of Coundon Gate Colliery, for the same reason.[11] The Black Boy branch was finally opened on 10 July 1827, from which time the compensatory payments to the aggrieved colliery owners were terminated.[12]

The joint owners of the Black Boy Colliery (with Joshua Ianson) were the Quaker Backhouse family of Darlington, staunch supporters of the S&DR. It is all the more surprising, therefore, to find that in 1839 Jonathan Backhouse contemplated using the rival Clarence Railway for his coal shipments, on account of the 'excessive' toll charges levied by the S&DR. Although this contentious alternative was not implemented at the time, the dispute came to a head in October 1848 when the colliery agent wrote to the S&DR Secretary, Oswald Gilkes, to inform him that henceforth Black Boy coal would travel over the Clarence Railway to be shipped from Hartlepool, and not the S&DR to be shipped from Middlesbrough.[13]

Some of the railwaymen who worked the Black Boy branch were housed by the S&DR at Rose Cottages, which still exist, opposite the site of the winding engine-house. They continue to bear the company's property plate, G 12. On the other side of the tracks to the engine-house, at the north end of Cheapside, Nicholas Downing's Phoenix iron and brass foundry was established because of the presence of the line. It was served by its own sidings to transport raw materials, fuel and finished products. Close by the southern end of Cheapside on the summit level an original chauldron wagon has been set up on a short length of fish-bellied rail. A stretch of about 100 yards of the line northwards has been marked out in the roadway and footpath. The later history of the Black Boy branch was greatly affected by the opening of the Shildon Tunnel Railway in 1842. It is dealt with in Chapter 8.

CROFT DEPOT BRANCH

Four years after the opening of the railway a rather longer branch was constructed as authorised in the Second S&DR Act of 23 May 1823. It diverged from the main line at Albert Hill in Darlington, 600 yards beyond the River Skerne crossing, and continued south for 3½ miles to terminate at a coal depot and wharf at Croft on the banks of the River Tees. The branch was opened on 27 October 1829 and 'large quantities of goods and minerals' began to be delivered into the North Riding of Yorkshire across Croft Bridge.[14] Hylton Dyer Longstaffe provides a colourful account of the opening day: '[The procession] consisted of numerous coaches, each drawn by one horse, crowded with from 30 to 50 passengers, and supplied with banners. These were followed by a train of waggons laden with coals from every different mine for the supply of the North Riding. The Company gave a cold collation at the Croft Spa Hotel. Mr Mewburn, their solicitor, was in the chair.'[15]

From the beginning, a horse-drawn passenger coach service was instituted. It was withdrawn on 13 December 1833. In the early years the coal wagons were also drawn by horses and not by steam power. Although the line was called the *Croft* Coal Depot branch, and the NER station on the east coast main line high above was also called *Croft*, in fact both the depot and the station were located in Hurworth Place, on the Durham side of the river. Croft Station had 'Spa' added to its name in 1896, but the community of Croft Spa lies just across the bridge on the Yorkshire side. The Yarm and the Croft Coal Depot branches had this much in common.

Haggerleases Branch

The proprietors' desire to proceed with a Haggerleases branch was the principal reason for their promotion of the Third S&DR Act, which received Royal Assent on 17 May 1824. It was titled 'An Act to authorise the Company of Proprietors of the Stockton and Darlington Railway to relinquish one of their Branch Railways and to enable them to make another Branch Railway in lieu thereof'. The relinquished branch referred to was No. 3 (above), from Norlees House to Evenwood Lane, authorised by the First S&DR Act. Although it was not proceeded with at the time, the Haggerleases branch substituted for it follows a very different alignment. This was chosen to suit both the Earl of Strathmore and the Revd William Luke Prattman, the owner of the Butterknowle and Copley Bent Collieries. Whereas Overton's original Evenwood Lane branch would have been a comparatively short one of 2.4 miles, commencing near the summit of the Etherley Ridge and descending to terminate near Evenwood Bridge on the River Gaunless, the Haggerleases branch as built, at 4.8 miles, was twice the length and followed the river throughout.

As before, a Plan and Sections were required to be deposited with the Clerk of the Peace in Durham. Although George Stephenson was still technically the full-time engineer to the line, he entrusted the survey to his son Robert, whose name duly appears on the title of the Plan and Sections, which were deposited on 27 September 1823: 'Plan of the Railway or Tramroad from the River Tees at Stockton to Witton Park Colliery and of the several branches therefrom all in the County of Durham respectively authorised to be made by two several Acts of Parliament passed in the 2nd and 4th Years of the Reign of his present Majesty King George IVth and also a Plan and Sections of several New or Additional Branch Railways or Tramroads proposed to be made from the said Main Railway 1823 Robt Stephenson Engineer J Dixon Surveyor.'[16] The 'several New or Additional Branch Railways' were the Haggerleases branch and three loop lines that enabled the Croft and Yarm branches to have more convenient junctions with the main line: an eastern loop for the Croft branch, and eastern and western loops for the Yarm branch. It says much for George Stephenson's paternal generosity of spirit, and the confidence he reposed in his son, that he was implicitly prepared for Robert to take the credit for the main line also, which George had surveyed and which appeared on his own Plan, deposited on 28 September 1822. Nevertheless, surveying for the Haggerleases branch and the loop lines, as shown on George Stephenson's Plan, were substantial pieces of work, competently undertaken. The survey of the Haggerleases branch can certainly be regarded as Robert Stephenson's first separate railway engineering commission.

The account Robert Stephenson subsequently submitted to the company for his services, amounting to £71 9s 0d, includes 15 guineas for 'setting out, levelling and making Section of Haggar Leases Branch Railway'. It also includes an item: 'To 34 Days attendance in London when the above branch was before Parliament – £35 14s',[17] which indicates that the S&DR proprietors had sufficient confidence in Robert Stephenson's abilities to entrust him with the responsible task of representing their interests before Parliament. In L.T.C. Rolt's words: '[The account] records Robert Stephenson's debut as a railway engineer on his own account, and proves that at twenty-one he was not only capable of carrying out a survey unaided, but also possessed the self-confidence and ability required to shepherd the result of his work through Parliament.'[18] (It may be that, because the engineer was expected to give time-consuming evidence on the railway in the House of Commons, George Stephenson, who had neither the time nor the inclination after his earlier bruising experience there, was content to give his son the credit). Robert Stephenson submitted his account to the company during May 1824. On the 18th of the following month, June 1824, he sailed from Liverpool for South America.

As built, this branch diverged from the main line immediately to the north of its crossing of the River Gaunless at St Helen's Auckland – marked on the Plan as beginning 'near School Door at St Helen's Auckland'. It then followed the river westwards by way of Ramshaw (Evenwood Lane), and Low Lands for 4.8 miles to terminate at West Mill on Haggerleases Lane, just beyond 'Butter Knowle'.

Although work on the branch commenced soon after authorisation, when only the first half-mile from St Helen's Auckland had been completed it was suspended for some years, again because of the company's 'liquidity crisis'. It was the Fourth S&DR Act of 23 May 1828 (Act 9 Geo IV cap. 60) that provided for the Middlesbrough Extension, which also compelled the S&DR to begin construction of the remainder of the branch within three months, and to complete it within three years. This was a concession to the Revd Prattman, who had threatened opposition to the company's eastwards extension on account of the continued isolation of his Butterknowle and Copley Bent Collieries from a rail terminus. Accordingly, work recommenced in July 1828 and the branch opened as far as Cockfield Fell on 1 May 1830 with a train from Darlington. It was completed and opened for traffic throughout on 1 October 1830. By 1848 it was serving no less than twelve collieries connected by private lines.[19]

Because of the nature of the terrain, a quarter of a mile from the terminus at Haggerleases Lane the branch had to cross the River Gaunless from the north to the south bank on what is believed to be only the second skew-arch on a railway in Britain. (The first was built by George Stephenson at Rainhill on the L&MR.) The skew-arch was built at an angle of 27 degrees. It is believed that, because of the untried nature of the engineering techniques, a full-size prototype was built from timber in an adjacent field before the construction of the stone arch commenced. The bridge was designed by Thomas Storey of the S&DR and built by James Wilson of Pontefract. It was opened in 1830, remaining in use until 30 September 1963. It still stands today, 12 ft wide and 42 ft long to the ends of the parapets, which are embellished with circular piers. It is impressive in its simplicity, the courses of large limestone blocks adorned only with string courses to mark the curve of the arch and the level of the roadway, and a chamfered coping on the enclosing walls.[20]

The sub-committee's minutes for 8 May 1830 describe this bridge as a 'swin bridge', the Durham dialect word for 'skew' or 'diagonal'. In 1857 the first OS surveyors mistook this for 'swing' and accordingly, but highly inappropriately, labelled it 'swing bridge' on their maps. A shorter contemporary bridge of similar appearance but conventional design finally carried the branch over the River Gaunless to its terminus immediately beyond on the northern bank. Most of this bridge also still remains *in situ*.[21]

There was a rather infrequent horse-drawn passenger service to 'The Lands', a station 1.4 miles short of the terminus at Haggerleases Lane. This service was an extension of the New Shildon to St Helen's Auckland service over the Brusselton inclines, which the company took over on 1 December 1833. A timetable dated 15 March 1836 indicates that a train to Darlington left Lands at 6.15 a.m. and returned from Darlington at 5.30 p.m. Two years later, the timetable of 20 March 1838 advertises 'Additional Conveyances – a Railway coach [*sic*] to the west station, near Cockfield, may be engaged on application to the guard [of the Darlington to St Helen's train], who accompanies the coach to and from St Helen's Auckland'. The west station was probably at Haggerleases Lane, the terminus, and the public timetables continued to include Haggerleases until August 1859. The branch continued to be worked by horse power until 1856 when locomotives took over. Only then was horse traction completely phased out over the entire S&DR system. The passenger service was finally withdrawn in 1872.

The Middlesbrough Extension

It is clear from the terms of its first prospectus that, from the beginning, the S&DR anticipated revenue from an increase of trade between Durham and North Yorkshire. This was the rationale that prompted the company to seek approval in the Acts of 1821 and 1823 for the building of the Yarm and Croft branches, respectively, from the main line, even though both stopped short of a crossing of the River Tees. At this time, in this part of the North Riding, there was no industry apart from agriculture. The branches were principally intended to serve the limited needs of this community – to transport consumer goods inward and produce outward. In contrast, the first rail crossing of the Tees much further east was dictated by the far greater potential for revenue from a very different kind of trade.

The navigable River Tees provides a classic example of a historic process in which the principal port moves inexorably downstream over the course of time. Not long after the opening of the S&DR, it became clear that the coaling staithes at Stockton were unable to cope with large vessels, which were also hampered by shoals in the river, notably at Jenny Mill's Island. The Yarm and Stockton parties of the S&DR favoured the Tees Navigation Company's proposal for a second 'cut' in the River Tees at Portrack, and other new works. The Darlington party favoured an extension of the line to new facilities to be built further down-river. In the event, the Darlington party won the vote, and proposals for what became the Middlesbrough Extension Railway began to take shape.

Two alternatives were available to the proprietors of the S&DR. They could either extend to Haverton Hill on the northern side of the river, or else to Middlesbrough on the southern side. The Haverton Hill site had become 'vacant' as a consequence of the Parliamentary rejection of Christopher Tennant's proposal for a direct line to the Auckland coalfield, the Tees and Weardale Railway scheme, for which it was to provide the terminus. Both routes were surveyed by Thomas Storey and Richard Otley. The Middlesbrough alternative was chosen because it provided the shortest and cheapest route. George Stephenson, appointed consulting engineer, made a preliminary survey in September 1827 for the extension to a new terminus 'in a close adjoining the River Tees, in the township of Leventhorpe or Middlesbrough'. Stephenson's favourable report was received on 22 September 1827 and the company was persuaded to instruct J.U. Rastrick to carry out a detailed survey. Estimates were then prepared by the company's own resident engineer, Thomas Storey, which were accepted at a general meeting of the S&DR on 19 October 1827. At a meeting on 5 January the following year, a successful resolution was proposed by Edward Pease and seconded by Jonathan Backhouse that 'the company should proceed with [the] railway extension whilst maintaining friendly relations with the navigation company'. It was this resolution that prompted three pioneer supporters of the S&DR to resign from the management committee – Leonard Raisbeck of Stockton, co-solicitor with Francis Mewburn, the Chairman, Thomas Meynell, of Yarm, and Benjamin Flounders, also of Yarm. Despite strong opposition in the House of Lords, the extension was duly authorised by the Fourth S&DR Act of 23 May 1828. It was simply titled 'An Act authorising the Company to make a new line from a Junction at Bowesfield, Stockton-on-Tees, across the River Tees, to Middlesbrough'.

The next task was the acquisition of the land necessary for the proposed new coal port. On 18 August 1828 Joseph Pease, who had been made responsible for the new development, recorded in his diary:

> Took a boat and entering the Tees Mouth sailed up to Middlesbrough to take a view of the proposed termination to the contemplated extension of the railway and was much pleased with the place altogether.

Its adaptation to the purpose far exceeded any anticipation I had formed – imagination here has ample scope in fancying a coming day when the bare fields we then were traversing will be covered with a busy multitude and numerous vessels crowding to these banks denote a busy seaport.

Who that has considered the nature of British enterprise, commerce and industry will pretend to take his stand on this spot and pointing the fingers of scorn at these visions exclaim 'that will never be'? If such an one appears, he and I are at issue. I believe it will be.[22]

Prophetic words!

At this point a small group of Quaker bankers and financiers – Joseph Pease and Edward Pease Junior of Darlington, Thomas Richardson of Great Ayton, Joseph Gurney, Henry Birkbeck, Simon Martin of Norwich, and Francis Gibson of Saffron Walden – independently formed a company that they called the 'Owners of the Middlesbrough Estate' to develop the site.[23] Surveys and negotiations with the landowner William Chilton then took place and finally 521 acres were purchased in May 1829 for £30,000, to accommodate the proposed coaling staithes on the riverside and, as we shall see, an entirely new township. A design competition, with a prize of 150 guineas, was held for the new staithes. It was won by Timothy Hackworth with an ingenious arrangement that enabled wagons to unload directly into the holds of ships moored below. The design also allowed for coals to be discharged without undue breakage – an important factor as the market value of coal depended on its size.

Altogether, six steam-powered staithes of this type were built. On 27 December 1831 the first cargoes were loaded on board the brig *Sunniside*, witnessed by the proprietors who then, with their guests, sat down to a banquet held in the long gallery of the staithes, Francis Mewburn presiding. The load of coal brought down included 'an immense entire coal weighing upwards of two tons' from the Black Boy Colliery[24] which was placed on the brig *Maria*, leaving for London the following day.

To return to the construction of the Middlesbrough Extension Railway, which proceeded apace. Its malleable iron rails were heavier than those of the main line – 33 lb per yard compared to 28 lb – fastened to the chairs by iron wedges on the 'keyed' plan and resting on oak sleeper blocks. The line was opened on 27 December 1830, the inaugural train being

---

TIMOTHY HACKWORTH'S COALING STAITHES

The ingenuity of the design lay chiefly in the method of unloading. Having been run into a long shed the wagons were raised 18 ft by a steam engine to a covered-in gallery 450 yards long which ran the whole length of the staithes. Horses then pulled the wagons to the drops over the hatchways of the ships. A single wagon was placed in a cradle and lowered to the deck of the vessel. A man rode down with the hopper wagon and released a bolt on the bottom and the coal was thus discharged directly into the hold. A counter-weight, which had been raised by the descending full load, then reasserted itself and lifted the empty wagon back to the level of the upper rails, at the same time bringing another loaded wagon down to the ship. The empty wagon was then pushed to the opposite end of the platform to its entry and lowered to the rails at right angles, again by steam power. It was placed on the rails with a 'twisting and jerking motion' which impelled it along the line about 100 yards to the sidings.

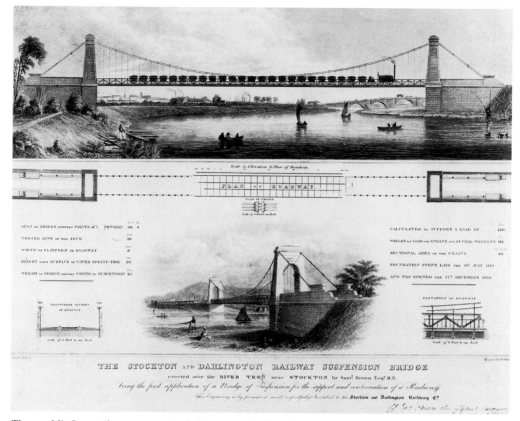

The world's first railway suspension bridge over the River Tees near Stockton, on the Middlesbrough Extension Railway. (Science Museum/Science & Society)

hauled by Timothy Hackworth's 0–4–0 No. 9 *The Globe*.[25] At the same time, the riverside terminus was named Port Darlington. The remaining 'Stockton Party' constituents of the S&DR must have found this a bitter pill to swallow, no doubt regarding it as a deliberate affront, especially since it was clear that Stockton's trade and importance would inevitably decline as that of Port Darlington increased. Perhaps in deference to their feelings, Port Darlington was dropped after a few months in favour of Middlesbrough.

The 4 mile extension commenced at Bowesfield junction, four-fifths of a mile short of St John's crossing near the original terminus. Almost immediately, it was required to cross the River Tees without causing even a temporary obstruction to the navigation – the Tees Navigation Company perversely stipulated that the river should not be obstructed by piers. Preliminary plans for a cast-iron bridge were drawn up by the S&DR's surveyor James Dixon. Although he proposed an ingenious method of construction, which was later employed by George Stephenson, and by Isambard Kingdom Brunel in the building of the Royal Albert Bridge at Saltash, his plans were rejected. Similarly, Timothy Hackworth's design for a wrought-iron girder bridge was rejected, partly because it was an untried technique – an ironic decision in the light of what was to follow. In the event, the S&DR appointed Captain Samuel Brown RN (1776–1852), who designed what became the world's first railway suspension bridge, and the first railway bridge over a navigable river, at a cost of £2,300. The bridge was

Watercolour, *The Opening of the Stockton and Darlington Railway*, by the Darlington artist John Dobbin (1815–88), which was commissioned for the jubilee celebrations of the S&DR in 1875. (Courtesy of Darlington Borough Council)

Edward Pease, 'Father of Railways', portrait by Heywood Hardy (1842–1933). The painting was commissioned by Pease & Partners to hang in their Boardroom in Darlington. (Courtesy of Darlington Borough Council)

Joseph Pease, portrait by James MacBeth (1847–91). The portrait was 'Presented by Subscription to the Corporation of Darlington, his native town'.

Below: Robert Stephenson, portrait by John Lucas. In the background is the Menai Straits between Anglesey and mainland Wales, spanned by Stephenson's Britannia railway bridge. (National Railway Museum/Science & Society)

Above: A quasi-historical, allegorical portrait of four generations of the family of George Stephenson at Killingworth, by John Lucas (1807–74), titled *The Birthplace of the Locomotive*. George Stephenson, seated, demonstrates his patent safety lamp to his son Robert on the right. Robert Stephenson senior, George's father, in the middle distance, points to a 'Killingworth Locomotive'. Beside him stands his wife Mabel, (née Carr). Seated nearby is George's first wife Frances (née Henderson) with her barefoot daughter who, in reality, died aged three months. George's second wife, Elizabeth (née Hindmarsh) is seated centre foreground, reading the *London Journal* of September 1848 in which her husband's portrait and obituary appeared. This painting hangs in the Boardroom of the Science Museum in London. Another version (probably the original) hangs in the Town Hall, Chesterfield. (National Railway Museum/Science & Society)

Richard Trevithick by John Linnell (1792–1874). In the background is a Welsh mining valley, scene of his early labours. (Science Museum/Science & Society)

George Hudson, 'The Railway King', as Lord Mayor of York, painted by Sir Francis Grant RA and presented to the City Council in August 1848. (Courtesy of York City Council)

Locomotion No. 1 races a stagecoach during the inaugural run of the S&DR. (From a painting by Terence Cuneo, 1947. Courtesy of the Science Museum)

Portrait from life of Timothy Hackworth. (Artist unknown. Courtesy of Sedgefield District Council)

Richard Trevithick's Pen-y-Daren locomotive of 1804 and the *Cornwall* locomotive of 1847. (From a painting by Terence Cuneo. Science Museum/Science & Society)

High Street House, Wylam, the birthplace of George Stephenson. In the foreground is William Hedley's locomotive *Puffing Billy*. (From a painting by Ronald Embleton. Courtesy of Frank Graham Ltd)

Oil painting of the inaugural train crossing Skerne Bridge, Darlington. (Artist unknown. Courtesy of the Stephenson Locomotive Society)

Locomotion No. 1 hauling a train of chauldron wagons past the booking office near the Stockton terminus of the S&DR. (From a painting by John Wigston. Courtesy of Frank Graham Ltd)

The King's Head Inn, Northgate, Darlington. Next door, at the beginning of Prebend Row, is 'The Great North of England, Newcastle and Darlington Railway Company's Parcels Office'. (From a painting by William Dresser. Courtesy of Darlington Borough Council)

Etherley Incline in 1875. (From a watercolour, artist unknown. Courtesy of Sedgefield District Council)

Brusselton Incline and engine-house, with Timothy Hackworth's patent winding gear, in 1875. (From a watercolour, artist unknown. Courtesy of Sedgefield District Council)

The poster produced by the L&NER to mark the centenary celebrations of the opening of the Stockton & Darlington Railway. (Designed by Andrew Johnson. National Railway Museum/Science & Society)

412 ft in length overall, from Peel Nook Stockton to Carr House Field Thornaby. Its main span was 281 ft long, 16 ft wide, and 20 ft above the high spring tide level. The decking was supported by 110 perpendicular rods hung from a total of twelve chains, six on each side.[26]

At a mere 111 tons the bridge was exceptionally light, and yet the S&DR had been led to believe that it would be capable of carrying weights of up to 150 tons. Not surprisingly, successive tests involving a weight of 18 tons placed in the centre, and trains of 45 and 68½ tons crossing the bridge, resulted in alarming deflections and some damage. Hylton Dyer Longstaffe gives a graphic account of the results of the tests: 'The first trial was made with sixteen [wagons], upon which the bridge gave way, i.e. as the sixteen carriages advanced upon the platform, the latter, yielding at first to their weight, became elevated in the middle, so as by degrees to form an apex, which was no sooner surmounted by half the number than the couplings broke asunder, and eight carriages rolled one way, and eight another – the one set onward on their way, and the other back again.'[27] Years later Robert Stephenson wrote: 'Immediately on opening the suspension bridge for railway traffic, the undulations into which the roadway was thrown, by the inevitable unequal distribution of the weight of the train upon it, were such as to threaten the instant downfall of the whole structure.'

A partial solution to the problem was eventually found by connecting batches of four wagons together by means of chains and couplings, which kept them 27 ft apart. Nevertheless, the bridge continued to prove unstable, especially in high winds. A tale is told that one of the drivers, being suspicious of the bridge, at first refused to cross with his locomotive. Accordingly, he reduced the speed of the train to a crawl as he approached the bridge, alighted, and ran ahead of the train, to climb back on the engine again after it had passed safely across.

After only fourteen years of unsatisfactory use the bridge was replaced in 1844 by a more conventional one designed by Robert Stephenson.[28] This had five cast-iron girders (three river spans of 89 ft each and two land spans of 31 ft each) supported by masonry piers. That bridge was itself renewed in 1907, the piers and foundations of the 1844 structure being re-used. The opening of the extension was commemorated with the issue of what was historically the second railway medal – the first had marked the opening of the L&MR in 1830.

An important development that followed the opening of the Middlesbrough Extension Railway was the establishment at Middlesbrough of a repair shop for locomotives – significantly

The medal issued to commemorate the opening of the Middlesbrough branch on 27 December 1830. (Courtesy of the National Railway Museum)

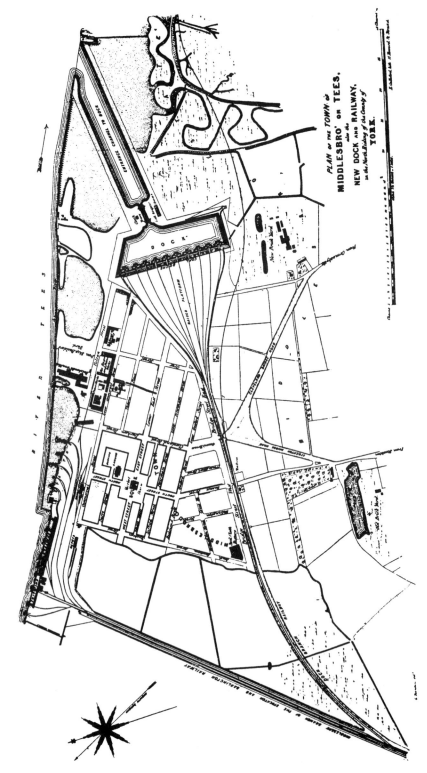

Plan of the town of Middlesbrough on Tees with the new dock and railway, as reproduced in The North Eastern Railway by W.W. Tomlinson.

at the other end of the main line from the major repair facilities at Shildon. This was an essential requirement in view of the expansion of traffic and the company's policy of exclusive reliance on locomotive haulage on the main network. In 1842 the Quaker Isaac Wilson (a member of the S&DR management committee) joined forces with fellow Quaker Edgar Gilkes (formerly an engineer at Shildon) to lease the Middlesbrough repair workshops under the name Gilkes, Wilson and Company. In 1844 it changed its title to the Tees Engine Works, manufacturing, maintaining and repairing locomotives both for the S&DR and other railways. From the 1850s onwards, the firm was the main builder of locomotives for the S&DR.[29] In addition, it manufactured engines and equipment for ironworks and mines, blast furnaces and agriculture.[30]

Another, perhaps not unexpected, consequence of the opening of the Middlesbrough Extension Railway was its effect on the price of coal. The opening of the S&DR in 1825, and its anticipated effect on local coal prices, marked the beginning of a process. However, the impact of the innovation of transporting coal by rail was comparatively limited until the opening of the Middlesbrough Extension line in 1830. Whereas the S&DR carried 46,216 tons of coal in the operating year 1828/9, in 1830/1 it carried 152,262 tons and in 1832/3, 336,000 tons. With this increase in volume, the economics of scale dictated that the local price of coal fell dramatically.

The first Middlesbrough Station was simply a coach-shed, opened on 7 April 1824, which was sited between 1824 and 1837 on a siding near the river and the coal staithes. The shed was moved in 1837 to a new site on the north side of Commercial Street where, between Stockton Street and North Street, it was replaced by a larger passenger station. Because of the poor state of the road from Stockton to Middlesbrough it became common practice to walk along the railway track. The company stationed a policeman at the old river bridge to prevent this. The Stockton Police Court records for 1841 show that quite a number of people paid 6*d* to travel on the engine of the coal train – after the departure of the last passenger train – to avoid prosecution for trespass. For the benefit of those who could not afford the normal 4*d* fare from Stockton to Middlesbrough, and also to secure additional revenue, the company introduced a third-class fare of 2*d* on 14 December 1835. Even when locomotives were the rule for hauling passenger trains, an exception was made on Sundays when one horse-drawn coach sufficed.[31]

MIDDLESBROUGH DOCK

By 1838 not only were the facilities at Port Darlington proving inadequate in their turn, but developments of the dock facilities at West Hartlepool were causing concern to the proprietors of the S&DR and the Owners of the Middlesbrough Estate. The very existence of Middlesbrough new town, the coal staithes at Port Darlington and the railway itself were threatened by the spread of the railways to the coast at West Hartlepool, where deeper waters were available on all tides. On 7 August 1839 the S&DR management committee noted that in the light of the increasing trend of coal shipments, new drops and staithes were required.[32] Accordingly, they decided to collaborate with Thomas Richardson, Joseph Pease and Henry Birkbeck, acting in the name of the Owners of the Middlesbrough Estate, in the construction of a new dock half a mile to the south-east of Middlesbrough 'to keep pace with the improvements of adjacent ports'.[33] (This was a clear reference to the competition posed by the growth of Hartlepool and the proposed Stockton and Hartlepool Railway.)

The new dock was designed by William Cubitt. Construction began in 1840, and was completed in May 1842. It was an ambitious undertaking. 'Middlesbrough Dock' was free from the effects of the tide, covered 9 acres (the water area being about 6 acres), and accommodated 150 ships and 1,200 loaded wagons. The depth at spring tide was 19 ft. It was approached by a 1.2 mile branch line from Old Town Junction, just short of Port Darlington, which terminated in ten diverging double lines, on a triangular raised platform covering 15 acres, each pair of lines serving a coal staithe on the west side of the dock. The ten staithes were together capable of loading ships at the rate of 105 tons per hour.[34] This short branch line was undertaken by the Owners of the Middlesbrough Estate, at their own expense, without benefit of an Act of Parliament, and was opened on 12 May 1842. The total cost of construction of the dock and branch line was £140,000, shared between the S&DR and the Owners of the Middlesbrough Estate.

Under an agreement of 1841 drawn up by Francis Mewburn, the S&DR's solicitor, the company took out a 999-year lease of the dock at an annual rent of 6 per cent of the construction costs.[35] The agreement allowed for outright purchase at some future date, and by the Consolidation Act of 1849 the S&DR was authorised to take over the dock from the Owners of the Middlesbrough Estate. (It was this Act that also made provision for the re-incorporation of the company.)[36] The dock passed into the control of the Tees Conservancy Commissioners when that body was formed under Admiralty auspices in 1852. (Most of its directors were also on the board of the S&DR.)

Middlesbrough Town

The Owners of the Middlesbrough Estate company was well named as they proceeded to build an entirely new town, Middlesbrough, while Port Darlington was being developed in 1830. The Owners commissioned Richard Otley to put their plans for a model town – 'an urban community free of squalor and disease' – into effect on the remainder of the site of 521 acres to the south and east of the port. They believed it should support a population of about 5,000, with four wide main streets (rather unimaginatively named North, South, East and West Streets) radiating from a central square, which later became the site of the first Town Hall; 123 building plots were developed, and the first houses were ready for occupation in April that same year, 1830. The new town reached its planned size within ten years. 'Such were the modest beginnings of the modern town of Middlesbrough, the first large settlement to be created as a direct result of railway development.'[37] Middlesbrough was also the first new industrial town to be laid out on a gridiron pattern of streets, the prototype for town planning in the decades that followed.

After the discovery of iron ore in the Cleveland Hills in 1850, and the development of the iron industry, Middlesbrough mushroomed to occupy an area far beyond the original site between Port Darlington and the dock, the river and the railway. 'It changed from a small coal town to a booming iron town virtually overnight. As men flocked to the ironworks here the population grew dramatically until old Middlesbrough began to burst at the seams.' The growth of Middlesbrough was indeed phenomenal. Whereas in the census of 1801 the total population of this fishing community was 25, and still only 150 in 1831, after the coming of the railway a mere decade later in 1841, it had risen to 5,463. The town's labour force then numbered 2,199 men, women and children. As recorded in the following three censuses the population expanded to 7,432 in 1851; 19,416 in 1861; and 39,563 in 1871. In 1891 it was a staggering 75,516.

Middlesbrough as it was in 1846 has been described as 'one of the commercial prodigies of the nineteenth century', as reflected by its 'proud array of ships, docks, warehouses, churches, foundries, wharves etc.'.[38] On a visit to the town in 1861, W.E. Gladstone was inspired to declare that it was 'the youngest child of England's enterprise . . . it is an infant, but an infant Hercules'.[39] It has to be said, however, that the founders' vision of a 'model' community was never realised, largely because of the unbridled increase in population, and the cramped and squalid conditions that followed in its wake.

To the Seaside

In step with the development of Middlesbrough, the proprietors of the S&DR shrewdly promoted the Middlesbrough and Redcar Railway Company, to serve the new town and the infant seaside resort of Redcar a further 8 miles to the east. It opened up the prospect of recreational travel along the coast of north-east Yorkshire. The 7.4 mile extension was authorised by an Act of 21 July 1845 and it was opened less than a year later on 4 June 1846. The ceremonial first train, consisting of a carriage and two trucks, was hauled by *Locomotion*.[40] The line commenced at Middlesbrough Dock Junction and terminated in the centre of Redcar near the present clock-tower. Soon after the opening, the company began holiday excursions for its own employees and their families, transporting 1,016 passengers in 1848 and 2,200 in 1850.[41] Although the Middlesbrough and Redcar Railway was nominally an independent company, it was at first worked by the S&DR and then leased by them with effect from 1 October 1847 for 999 years.

A third station, to serve the new town of Middlesbrough, was built at the foot of Sussex Street, just to the east of Middlesbrough Dock Junction. It was opened on the same day as the extension – 4 June 1846 – replacing the station in Commercial Street which became a goods depot. The new station proved to be a profitable investment. In December 1857 Joseph Pease noted in his diary, with some astonishment, that in 1851 the S&DR had booked 61,319 passengers at Middlesbrough: in 1854 the number was 89,679 and in 1857 it had reached 109,577.[42] The present Middlesbrough Station (the fourth!), in an impressive Gothic style, was the work of the architect William Peachey. It is located just to the east of the 1846 station and was opened on 1 December 1877.

Some 5 miles down the Yorkshire coast from Redcar was a small fishing community called Saltburn, at the mouth of Skelton Beck. After the success of Redcar, the Quaker proprietors of the S&DR realised the potential for development of this site as another genteel seaside holiday resort. They accordingly formed the Saltburn Improvement Company, after the precedent of the Owners of the Middlesbrough Estate, as the development agency of the proposed resort, and commissioned the Darlington architect George Dickinson to visit a cross-section of existing holiday resorts in preparation for drawing up suitable plans. According to his biographer, Henry Pease was the originator of this attempt to create a 'northern Brighton'. After walking along the coast to Saltburn in 1859, 'he had seen in a sort of prophetic vision on the cliff before him, a town arise, and the quiet unfrequented glen, through which the brook made its way to the sea, turned into a lovely garden'.[43] George Dickinson's recommendations fully accorded with Henry Pease's vision, and the building of the new resort proceeded apace.

This was to be the first, and almost unique, occasion when a railway company single-handedly founded a seaside resort. The names of the company's members are significant: Thomas Meynell, of Yarm; John Pease – Joseph's elder brother; Joseph Whitwell Pease – Joseph's son; Isaac Wilson, a Quaker, ironmaster, proprietor of the Middlesbrough Pottery, and Chairman of the Tees Conservancy Commission; Alfred Kitching; and Thomas McNay, agent for the S&DR. For the S&DR to reap the benefits of its enterprise it was necessary to extend the main line yet

The S&DR's Zetland Hotel, Saltburn-on-Sea, as it appears today.

again (but in the event for the last time), from Redcar to the new resort. The proprietors had another motive to extend their line to Saltburn. Among them, the dominant Pease family and their partners owned ironstone mines in East Cleveland, for which a line to Saltburn would prove an effective outlet once further extensions were completed. Authorisation was obtained in an Act of 23 July 1858, which also authorised the amalgamation of the Middlesbrough and Redcar Railway and the S&DR – hence its title 'The Stockton & Darlington (North Riding) Railway'. The line was opened from Redcar Junction to Saltburn, a distance of 5.6 miles, on 17 August 1861. At the same time, a new Redcar Station was built on the extended line, making Redcar Old Station to the north redundant. That station and the short stretch of line to it from Redcar Junction (which had become a spur) were accordingly abandoned.

At the Saltburn terminus a station was built on the heights overlooking the confluence of Skelton Beck with the sea. The Act of 23 July 1859 also included powers to build a hotel. A site was chosen immediately to the east of the station, with commanding views out to sea. The architect, who also designed the station, was William Peachey. The foundation stone of his Italianate creation was laid by Lord Zetland, after whom the hotel was named, on 2 October 1861. When it was opened, with thirty-seven bedrooms, on 27 July 1863, it was described as 'the finest and largest railway hotel in the world, with the facilities that some of its clientele would have expected in a country mansion'. To carry fuel to the hotel a siding was extended to its rear from the station. Shortly after, passengers could board their trains in London and be transported all the way, and actually into their seaside hotel under a glazed covered way, by means of this siding, without once having set foot on the ground. The Zetland may fairly claim to be the first railway hotel. It was sold in 1975 and the new, private owners have converted the accommodation into self-contained apartments.

With the spread of the railway network, attractive coast and countryside became accessible to the general public for the first time. During the lifetime of the S&DR, travel by rail for pleasure and leisure, for holidays and excursions, increased to such an extent that it was contributing a significant proportion of the company's revenues. It was policy then for the S&DR to promote both its passenger services, and resorts such as Saltburn as holiday destinations. It is a commentary on the times that while, during the period under review, freight – mineral traffic – had the priority, today passenger traffic is in the ascendancy.

We have dealt at length with developments east of Stockton because these 'extensions', 13.3 miles in total, were truly a *continuation* of the historic 26.5 mile S&DR main line between Witton Park Colliery and Stockton Quay, inaugurated on 27 September 1825. It is a remarkable fact that there is still today, 175 years later, a regular daily service over the 31.8 miles between Bishop Auckland and Saltburn via Darlington Bank Top Station, 28.4 miles of which traverses that original line (8.4 miles between Shildon Junction and Parkgate Junction, Darlington, and 20 miles between Oak Tree Junction and Saltburn).

The opening day of the S&DR on 27 September 1825 inaugurated 28 miles of line. From that date, the company gradually expanded its empire. In the main it did so not in its own name but, as we have seen, in connection with the Middlesbrough Extension Railway, by promoting nominally independent railway companies whose directors and officers were S&DR place-men.

In addition, the proprietors of the S&DR often held major shareholdings through which they exercised control. Thus the literature habitually refers to newly 'floated', 'authorised' or 'incorporated' companies being 'operated by' or 'leased to' the S&DR, and ultimately being 'sold to', 'absorbed by', 'merged with', or 'amalgamated with' the S&DR. Although there are subtly different nuances of meaning within these categories of terms, the practical outcome was much the same.

In its heyday, the S&DR's domain extended from Consett in the north to Tebay in the south-west, and from Penrith in the west to Saltburn in the east. However, apart from the main line, almost the whole of the network consisted of three routes that converged on Darlington to a greater or lesser degree, and to these we now turn our attention.

CHAPTER 8

New Lines:
The Building of an Empire

'OPENED FOR MINERAL TRAFFIC'

Almost invariably, two successive dates are given in the literature for the opening of a new line: first, 'opened for mineral traffic', and later, 'opened for passenger traffic'. This order reflects the priority given by early railways to mineral traffic over passengers as a source of revenue. This was especially so for the S&DR from the beginning. It was seen by its proprietors as essentially a slow-moving mineral railway with passengers, and even the movement of other merchandise, as secondary considerations. This is in marked contrast to the long-distance trunk railways that followed, and even the near-contemporary but relatively short L&MR, which derived a substantial proportion of its revenue from the conveyance of merchandise and passengers.

It has become an axiom that the building of the main line of the S&DR from Witton Park to Stockton was market led – to transport coal (not passengers) from south-west Durham to markets further afield. It also had a subsidiary, local, and more domestic purpose, which was highlighted on the opening day. This was to establish coal depots for communities en route, at Darlington, Croft and Yarm. In the course of time, similar depots were established elsewhere on the network. For example, the 1861 OS map marks a 'Coal Depot' on the north side of the main line where it crossed Haughton Road in Darlington. In addition, there is some evidence that the S&DR permitted certain stationmasters to act as coal merchants for their own areas. The stationmaster purchased coal direct from the collieries by the wagon-load, had it weighed and bagged, and sold it to local customers as a perk to augment his wages from the company.

Of course, the S&DR quite happily transported anyone's coal. As it expanded within the coalfield that was its heartland, a spider's web of lines developed, linking the many pits and drift mines with the S&DR system. Most of these were privately built and owned by the colliery proprietors. During the lifetime of the S&DR, however, coal production was concentrated at fewer, larger collieries. Similarly, with the development of the iron industry in the region, especially on Teesside, and its demand for coke (a by-product of coal) as an essential raw material, coke ovens were established alongside strategically sited collieries. Again, fortuitously, the railways were on hand to provide a vital fuel transport artery, to the mutual financial benefit of these two inter-dependent industries.

Of almost equal importance to coal in the domestic, local economy, was limestone and its derivative quick-lime, for a number of manufactures as well as agriculture. It is significant that on the 1861 OS map, lime depots are given equal prominence with coal depots, for instance at the halfway point of the Darlington Coal Depot branch, at St John's Crossing, Stockton, and at the end of the Croft and Yarm Coal Depot branches. The story of lime parallels that of coal in other respects. Originally, it was derived from many local kilns situated near to as many local quarries in a band of limestone that stretched across the whole of northern England. It was a valuable and vital resource. Again, during the course of time, the production of lime was concentrated in larger units accessible from the railways.

Limestone in bulk was also transported by the S&DR from its earliest days, and increased dramatically in volume and revenue-earning capacity. Crushed limestone was in demand as a necessary flux in the nascent blast-furnaces. Paradoxically, one effect of the new railways was a great expansion of the iron industry, which repaid its benefactor through charges levied for the transport of all three of its basic raw materials, including limestone. In 1818 the company conservatively estimated that only 5,000 tons of limestone would be required at Yarm and 1,000 tons at Darlington per annum. Those figures were to be greatly exceeded: by 1850, the total amount transported had reached 66,706 tons; by 1855, 249,000 tons; and by 1868, it had passed the 500,000 tons mark – mainly because of the development of the iron industry on Teesside, of which more later. In the early years of this period, however, much of the limestone was destined for the Derwent Iron Company's works at Consett, which were founded in 1841. By 1846, it possessed fourteen blast furnaces and thirty-five coal and ironstone mines. It was then the largest iron-making concern in England and second only to the Dowlais Company of South Wales in the entire country. Over time, the S&DR and the Derwent Iron Company greatly influenced each other's fortunes.

When we add iron ore to limestone and coal, therefore, we have the trinity of minerals which made up the staple goods traffic that came to constitute the lifeblood of the S&DR's commerce. Indeed, it was the conveyance of these three commodities specifically which was incorporated into the preamble to the First S&DR Act to justify the building of the railway. Francis Mewburn made a perceptive entry in his diary for 1854:

> S&DR. This Railway seems only in the infancy of its prosperity. The iron furnaces of Middlesbrough, and those about to be erected at Darlington and Stockton, will require an enormous amount of coal and lime. Eight furnaces consume nearly half a million tons of coal per annum. When 50 are erected what a quantity of coal and lime must pass over this Railway, and add to them the manufactured iron and the iron ore from the west, the passenger and the general traffic.[1]

Before the advent of railways there existed many small, local ironworks and bloomeries in the north-east. They were usually associated with small local deposits of iron ore, in proximity to collieries, which over time became exhausted. The presence of these ironworks was one of the factors which favoured the establishment of a railway industry in Darlington in the first place. Paradoxically, and in reverse, once the railways were established, they tended to dictate the location of larger ironworks, alongside these 'iron roads' that were so essential for the transport of their raw materials and finished products. A good example is the ironworks at Tow Law, established in 1851. In Darlington, the South Durham Iron Works was established in 1854 on a site north-west of the S&DR Crossing. Appropriately, the South Durham Iron Works concentrated on the manufacture of iron rails. It was the forerunner of the Darlington Forge, second only to the North Road Shops as the largest employer in the borough. A rather smaller ironworks was established in 1864 at Dinsdale, 3½ miles to the east of the Crossing. (The South Durham Iron Works closed in 1931 and was dismantled in 1947.)

The constraints of economic geography dictate that major industries initially tend to be located at or near the source of one or more of their raw materials. In the case of the iron industry, while limestone was widespread in the north of England, a principal source was the abundant beds quarried in Weardale. Coal was abundant in the Northumberland and Durham coalfield, and to a lesser extent in the West Cumberland coalfield where the strata were difficult and relatively costly to mine. Extensive deposits of iron ore were discovered and mined in Cleveland in the east, and in the Furness District in the west. Enormous resources of haematite, a high-grade ore, were discovered at Park, north of Barrow-in-Furness, in 1850. The end result of the concentration of the iron industry, therefore, was the establishment of four large and very productive ironworks complexes at four sites in the north of England: 1. at Consett in North Durham, distant from good sources of iron ore but on the margin of the Northumberland and Durham coalfield; 2. at Workington in the middle of the West Cumberland coalfield, and relatively near the Furness ironstone deposits; 3. at Barrow-in-Furness and at Millom, where the position was reversed; 4. at Middlesbrough in the east, on the doorstep of the Cleveland ironstone deposits but relatively distant from the Durham coalfield.

The S&DR and its associated companies were crucial in determining the location of these four ironworks conglomerates on the one hand, and their continued production and prosperity on the other, since the works depended on the railways for the transport of their essential raw materials. Lines were extended, and routes developed, specifically for this purpose. The traffic was by no means all one-way. The railways also served to transport the finished products of the iron industry on their way to markets at home and abroad. It is perhaps symbolic that the rolling mills at Workington came to specialise in the production of rails for tracks at home and abroad.

While much of the mineral traffic by-passed Darlington at the centre, the wealth it generated certainly did not. It flowed there principally through the purchasing power of the resident Quaker proprietors, who frequently had a financial interest in the enterprises their railways served, and, not less, the purchasing power of the resident railwaymen.

We turn now to consider four lines that exemplify the contribution of the S&DR companies to the economy of the north of England in general, and their role in facilitating the iron industry in particular.

THE WEAR VALLEY COMPLEX

The core of this complex was the Bishop Auckland and Weardale Railway. However, the line ultimately became much longer than its modest title would suggest, reaching as far as Consett, 21½ miles from Bishop Auckland and 23 miles from Darlington. Two stages of a route from Darlington to Consett were already in place by 1843 – the S&DR's main line from North Road Station to New Shildon, and the Shildon Tunnel Railway. As its name suggests, the latter involved the building of a tunnel 1,225 yards long through the hitherto daunting obstacle of the 500 ft high magnesian limestone Shildon Ridge. It ran from a junction with the S&DR's main line at New Shildon for 2 miles to South Church, 1 mile short of Bishop Auckland. The line was promoted, without the blessing of Parliament, by the nominally independent Shildon Tunnel Company composed of three S&DR directors, Joseph Pease, Thomas Meynell and Henry Stobart. This was an early example of the S&DR practice of building a line in the name of certain directors of the company. If the enterprise failed, the railway was not damaged. The practice also had the advantage that, in cases where the directors between them owned all the land, the line could be built without Act of Parliament. The line through Shildon Tunnel was opened on 19 April 1842 and was worked from the beginning by the S&DR.

The railway companies managed their opening ceremonies with some style, as an account written in 1913 but culled from contemporary sources makes clear:

About 10-o-clock on the opening day, a procession was formed at the Cross Keys Inn, this being announced by minute guns. The Union Jack was carried at the head of the procession, and immediately behind this emblem walked the resident engineer, Mr Luke Wandless, with Mr Henry Brook, the principal contractor, on his right, and Mr Thomas Dennies, the principal bricklayer, on his left, and accompanied by Lord Prudhoe's brass band, and also taking part in it were the inspectors, sub-contractors and workmen, and others who had been connected with the work. Following the procession was a vast gathering of people. With banners waving they processed into the tunnel by the southern approach. The darkness was relieved by innumerable candles, and a platform had been erected at which the ceremony connected with the Opening took place. All being ready Mr Booth, the contractor, presented to Mr Wandless a silver trowel with the following inscription: 'Presented on the 10th January 1842 to Mr Luke Wandless, the Resident Engineer of the Shildon Tunnel by a few friends for the purpose of laying the last brick'. Mr Dennies then presented the 'last brick' which along with the trowel, had been carried in the procession, and after a short address Mr Wandless deposited the brick in line with its millions of companions, pointing it with Roman cement by means of the silver trowel. A bottle of wine having been sprinkled on the last brick by Mr Booth, he Christened it 'The Prince of Wales' amid great applause. Simultaneously cannon were fired in the open air and the band struck up the favourite tune 'Merrily Danced the Quaker's Wife'. The procession then passed through the tunnel, emerging at the north end. Dinner was provided for the workmen at six different [public] houses in Shildon. The principal function was at the Cross Keys where about fifty prominent gentlemen partook of the entertainment.[2]

Two items in this account call for further comment. Despite its ambiguity, the title 'Prince of Wales' was applied not to the last brick, but symbolically to the tunnel as a whole, since the future King Edward VII was born on 9 November 1841, only two months before the tunnel was opened on 10 January 1842. The Cross Keys public house was situated almost immediately above the tunnel in Cheapside. A contemporary cartoon-style print of the banquet for distinguished guests at this 'focal point of the entertainments' shows the band playing on a raised dais, which bears the Darlington coat of arms.

The opening of the Shildon Tunnel branch called for the building of a new station to serve New Shildon. It was situated at its junction with the S&DR main line, just to the east of the end of the Black Boy branch and the Surtees Railway. It was opened in that same year, 1842. Between 1837 and 1842 a booking-office service only operated from the Mason's Arms. The present Shildon Station occupies the site of its 1842 predecessor.[3]

The Shildon Tunnel Line and the Black Boy Branch

Shildon Tunnel passed almost beneath the summit winding engine-house of the Black Boy branch, whose alignment crossed from west to east of the new line above the tunnel. The opening of the Shildon Tunnel Railway in 1842 had profound consequences for the Black Boy branch, chiefly because it provided a much less costly and more convenient outlet for coal, through the tunnel, to the main line at Shildon Junction. Five new lines were built, four of which effected a junction with the Shildon Tunnel Railway. The exception was a short branch that diverged about halfway down the Black Boy northern incline and fell steeply to limekilns owned by Emmerson Murchamp and set into the embankment of the Shildon

Tunnel Railway at the point where it crossed Eldon Lane. This enabled the limestone and coke to be fed into the top of the kilns from a siding, and the calcined lime to be transported from the foot onto the Black Boy branch. The kilns opened on Christmas Day 1845 and ceased operation about 1880. Typical S&DR railwaymen's cottages built alongside this branch, fronting Eldon Lane, still survive as 'Railway Houses' terrace. They carried the S&DR property plate G 13.

To the south of the limekilns branch a steep connecting spur was built about 1850 from the Black Boy branch to the Shildon Tunnel Railway near Tunnel North Junction. Apart from providing an outlet for Black Boy coal,[4] it also acted as a relief line should the tunnel be blocked or the line otherwise obstructed. It was in use as late as 1908.

A second new line was built from a north-facing junction with the Shildon Tunnel Railway, about 1,300 yards from the north end of the tunnel. It proceeded eastwards and crossed the Black Boy branch on the level between the limekilns branch and the connecting spur. After half a mile it turned north-east and contoured along the southern slope of the Eldon Beck Valley to serve the Eldon South Durham 'New' Pit (1841–1931), and was later extended to the Old Eldon John Henry Pit (1864–1928). This 2 mile private colliery line was first called the Eldon Colliery branch but later took on the grandiose title the South Durham Railway.[5]

Between Machine Pit and Black Boy Colliery on the originally intended Black Boy branch a large new colliery was opened out at Auckland Park in 1864. The third of our new lines was built from a junction with the Black Boy (Gurney Pit) Colliery Railway to serve this new colliery and connect it, by means of a reversal, with the Shildon Tunnel Railway about 700 yards short of South Church.

The connecting line from the Black Boy branch to the Deanery and Adelaide Collieries to the west was effectively severed by the deep cutting that led to the north portal of Shildon Tunnel. It was replaced by the fourth of our new lines, which curved round to the north to effect a junction with the Shildon Tunnel Railway opposite the junction with the Eldon Colliery Railway to the east.

Remarkably, the final date of closure of the Black Boy branch is not known. Although the opening of the Shildon Tunnel Railway might have been expected to sound its death-knell, it survived against all the odds as a persistent reminder of the earliest years of the S&DR.

The Bishop Auckland and Weardale Railway was authorised by Act of Parliament as early as 15 July 1837. The intention was to connect New Shildon with Crook, 8.4 miles to the north (where the Pease family had extensive coal-mining interests) via Bishop Auckland. The Shildon Tunnel Railway represented the first stage of that enterprise. The 6.4 mile second stage from South Church to Crook was opened on 8 November 1843. From the beginning, it was operated by the S&DR and leased by them from 8 November 1843. The incentive to build an extension from Crook to reach Consett, a further 11 miles to the north-west as the crow flies, was the establishment there of a large ironworks by the Derwent Iron Company.[6]

In the event, the S&DR built the major part of the extension, the 11 mile so-called Weardale Extension Railway, without Parliamentary sanction, from Crook to a junction at Waskerley Park[7] with the Wear and Derwent Railway, which already had access to Consett. The Weardale Extension Railway, opened on 16 May 1845, was built in expectation of revenues arising from the transport of raw materials – lime, coal and iron ore – over S&DR metals. Although this was built as a way leave line for the Derwent Iron Company by the S&DR, under an agreement of late 1844, it was virtually bought out by a consortium of S&DR directors, Joseph Pease, Thomas Meynell and John Castell.[8] The Wear and Derwent Railway had started life as the grandly titled Stanhope and Tyne Railway (the S&TR), the proprietors engaging Robert Stephenson as consulting engineer and T.E. Harrison, one of his most able assistants, as resident engineer, in 1832.[9]

The summit level of the S&TR at Whiteleahead was 1,445 ft above sea level – the highest point on the S&DR system, and indeed the highest point on the NER after amalgamation. It was also the second highest summit of any railway in Britain, only exceeded by the Dalnaspidal summit on the Highland Railway. So, considering that much of the route was over high moorland, the S&TR was completed in a remarkably short space of time, and was opened from Stanhope in Weardale to South Shields on the estuary of the River Tyne, a distance of 32.4 miles, on 10 September 1834. From the summit the line descended for 2 miles to its terminus at Stanhope in the valley floor below, by two extremely steep self-acting inclines. These were operated by static steam engines located at the Weatherhill summit and halfway down, at Crawley 1,123 ft above sea level. From the foot of the inclines, at Lonehead Farmhouse, half a mile north of Stanhope township, lines radiated out to a 'smelt house', for Weardale lead, calcining kilns, for processing Weardale limestone from nearby quarries, and a coal depot, at 796 ft above sea level. All were built by the firm of Whitfield Gardiner, whose name was recorded on a sundial there.

The S&TR was a good example of the way in which the early railways drew on the experience and the technology of the colliery wagonways. When first completed, 10½ miles of the line were worked by horses, 11 miles by no less than nine static steam engines totalling 375 hp, 3 miles by self-acting inclined planes, and only 9.25 miles by locomotives: 'every form of motive-power then known'.[10]

After its formation, the S&TR had a complex history. The 10.8 mile western section (the Wear and Derwent Railway) from Stanhope to Carr House near Consett was closed in 1839, together with the kilns in Stanhope, because it proved too costly to work. On the initiative of Robert Stephenson, a new company was formed, known as the Pontop and South Shields Railway, to take over the old company's property and debts. It was duly incorporated by an Act of 23 May 1842. One of its first undertakings was to sell the western section to the Derwent Iron Company, who revived the route known as the Derwent Railway, and used it principally to carry lime from the reopened kilns in Stanhope to the ironworks in Consett.

By the time the S&DR built the Weardale Extension Railway it had already leased the Wear and Derwent Railway (*q.v.*) from the Derwent Iron Company with effect from 1 January 1845.[11] It did so mainly because its eastern end provided the last link in the chain to connect Bishop Auckland in the south with Consett in the north. The link ran from Waskerley Junction via the celebrated Nanny Mayor's self-acting incline, Rowley,[12] and the gorge of Howne's Gill.[13]

The Weardale Extension Railway and the Wear and Derwent Railway together became known, *unofficially*, as the Wear and Derwent Junction Railway from March 1845.[14] A passenger service was instituted between Stanhope and Crook on 1 September 1845, which picked up passengers at Waskerley from Howne's Gill via Nanny Mayor's incline. It was short-lived, however, and ceased late in 1846.

An Act of 16 July 1855 authorised the building of a 2.4 mile deviation line between Whitehall Junction on the Weardale Extension Railway and Burnhill Junction on the Wear and Derwent Railway. This replaced the roundabout diversion north-west to Waskerley and the reversal down Nanny Mayor's incline. The more direct route was opened for mineral traffic on 23 May 1859 and for passengers on 4 July 1859. The building of the deviation also led to the abandonment of the stretch between Waskerley Park and Burnhill Junction, on the Wear and Derwent Railway (including Nanny Mayor's incline), even though it obliged traffic between Stanhope and Consett to travel on another roundabout route via Whitehall Junction.

> ## SURMOUNTING HOWNE'S GILL GORGE
>
> At Howne's Gill – a dry ravine 160 ft deep and 800 ft wide – 'a most remarkable device' enabled the railway to negotiate this obstacle. At the top of the self-acting inclines the wagons were turned sideways by means of a turntable and transferred to special incline cradles, with front and rear wheels of unequal diameters, running on two sets of rails, the outer one of 7 ft and the inner one of 5 ft gauge. They were lowered down and hoisted up the two precipitous inclines, with gradients of 1 in 3 and 1 in 2.5 respectively, by a single 20 hp fixed engine at the bottom of the gorge which worked both inclines simultaneously. Also at the bottom was a rectangular platform with four turntables at the corners which enabled the wagons to be turned at right angles and transferred from the foot of one incline to the foot of the other.[15]

The hilly nature of the terrain between Bishop Auckland and Consett posed formidable obstacles to the building of these lines. Initially, some of the more severe gradients were negotiated by inclines. The three most notable differed from the arrangements at Etherley and Brusselton. Nanny Mayor's and Sunniside, between Crook and Tow Law, were single inclines, the latter 1¾ miles long with gradients as steep as 1 in 13 and 1 in 16. In the course of time, all these inclines were circumvented, enabling locomotives to take over throughout.

The Sunniside incline was avoided when a more circuitous but less severely graded line was built between Crook and Tow Law, which added 1.2 miles to this section.[16] It was authorised by an S&DR Act of 3 June 1862, but was completed and opened for mineral traffic by the NER on 10 April 1867. Finally, the Howne's Gill inclined planes were replaced by a mighty viaduct, 700 ft long, with 12 arches, 150 ft high in the centre. It was designed by Thomas Bouch, work commenced in March 1857, and it was opened on 1 July 1858. The viaduct was constructed of stone and over 2½ million firebricks from Peases' brickworks at Crook.[17]

The line from Howne's Gill to Consett was served by a pair of self-acting inclined planes powered by a stationary steam engine at the summit at Carr House. The S&DR built a short 1½ mile spur to Blackhill (later incorporated in Consett town) from a point just beyond the Howne's Gill viaduct. This enabled a through, timetabled passenger service to be introduced between Blackhill (for Consett) and Darlington, a distance of 30 miles. It was a remarkable achievement.

The first, but long-delayed objective of the Bishop Auckland and Weardale Railway was finally achieved with the passage of the Wear Valley Railway Act on 31 July 1845, promoted by the S&DR. The history of this railway (to which the western section of the S&TR was sold), well illustrates the convoluted and complex nature of railway enterprises in County Durham. It ran from the Wear Valley (Witton) Junction on the Bishop Auckland and Weardale Railway (of which it was effectively a branch) for 10¾ miles to Frosterley. Even before opening day, on 3 August 1847, this lusty infant swallowed up its parents when, by an Act of 22 July 1847, by which time it was leased by the S&DR, it was authorised to purchase or lease the Bishop Auckland and Weardale Railway, the Shildon Tunnel Railway, the Wear and Derwent Railway and the Weardale Extension Railway. Two months later, the S&DR invoked a section of the 1845 Wear Valley Railway Act, enabling it to take over the enlarged Wear Valley Railway. These two moves brought about the consolidation of the five associated railways into one workable unit, which was leased by the S&DR for 999 years from 1 October

1847, and wholly absorbed on 23 July 1858. That merely regularised what had been *de facto* from the beginning – control by the S&DR through nominally independent companies.

The pattern was repeated with the authorisation, in an Act of 28 June 1861, of the S&DR-sponsored Frosterley and Stanhope Railway, a 2½ mile extension of the Wear Valley Railway. The Act also authorised a short 2.4 mile branch, which diverged just short of Frosterley at Broad Wood and followed the Bollihope Burn south to serve immense limestone quarries at Bishopley Crag. It is an interesting sidelight on the S&DR's involvement with many of the industries it served that the company owned the Bishopley Quarries and the associated limekilns. For many years the work at both was let out on contract.

The Frosterley and Stanhope Railway was opened for mineral traffic as far as Newlandside on 30 April 1862, and to Stanhope Old Station for mineral and passenger traffic on 22 October. The line was absorbed by the S&DR by an Act of 30 June 1863 even before it was opened for passenger traffic, and it was operated by the S&DR from the beginning. The combined line from Wear Valley Junction to Stanhope was principally promoted in order to tap the considerable mineral wealth of Weardale. The railway benefited by transporting lime and lead, refined from Weardale's abundant reserves of limestone and lead ore, to markets further afield. The S&TR had performed the same function, but only to a limited extent because the locomotive-hauled line terminated at Weatherhill on the brink of the escarpment high above Stanhope. The cost of operating the inclines from Weatherhill to Stanhope was uneconomic and it is not surprising, therefore, that the whole line had already closed when the much easier outlet, the Frosterley and Stanhope Railway, was completed. Although the two systems were tantalisingly close, there was neither reason nor opportunity to connect them.

After the S&DR completed its remarkable takeover of the five constituent railways of the Wear Valley Network it was authorised by an Act of 16 July 1855 to built a 3.2 mile branch from Crook, at the northern end of the former Bishop Auckland and Weardale Railway, to High Hill Farm near Waterhouses in the Deerness Valley. When it was opened in 1858, it made an end-on junction with the Deerness Valley Railway from Durham City which was authorised only a fortnight after its own Act, on 30 July 1855. The Deerness Valley Railway was promoted by Joseph Pease and his son Joseph Whitwell Pease to serve their new colliery on the royalty of Lord Boyne at Waterhouses. It was worked from the outset by the NER. The principal purpose of the S&DR's branch from Crook, however, was to serve two collieries en route – Stanley and North Roddymoor. A short private branch line also served Wooley Colliery to the east. The Waterhouses branch incorporated two inclines, the southern climbing up from Crook for a mile at 1 in 16. Both inclines were worked by a static steam engine at their summits.

The completion of the Bishop Auckland and Weardale Railway had a marked and decisive effect on the historic Etherley inclines. Its route from Bishop Auckland followed the south bank of the River Wear as far as Etherley, before bridging the river and crossing to the north bank. It was a simple expedient to effect a connection between the Witton Park group of collieries and the new line, and then to haul the coal wagons by locomotive power via a comparatively level 'back door' route through Shildon Tunnel to Shildon Junction. From the opening of the Bishop Auckland and Weardale Railway on 8 November 1843, therefore, the Etherley inclines to West Auckland were made redundant.[18] The OS map of 1861, which was surveyed between 1855 and 1858, indicates that by that time the tracks of the Etherley inclines had been lifted.

The Brusselton inclines still continued in operation to serve the collieries of the Gaunless Valley and Brusselton Colliery itself, since they possessed no alternative outlet. In the S&DR Act of 3 July 1854, a 3.6 mile connecting line – the so-called Tunnel branch (although it did not in fact incorporate Shildon Tunnel) – from West Auckland Junction to Shildon Tunnel North Junction, via Fieldon Bridge, was authorised. This last link in the route from West Auckland to New Shildon was opened on 13 September 1856. At last, the formidable Shildon ridge was by-passed. This eventually led to the abandonment of the Brusselton inclines in 1858, together with a branch northwards, opened in 1830, from Brusselton west incline 'bankfoot' to St Helen's Colliery in the Gaunless Valley (later extended to Woodhouse Close Colliery). The by-passing of the inclines came none too soon, for with rising traffic levels after 1850 they had developed into a major bottleneck, and all the more so as the network to the west of Shildon contained the last vestiges of horse-drawn operations on the whole of the S&DR system. Two years after the opening of the Tunnel branch, a passenger service was inaugurated, and the service over the Brusselton inclines, which had begun as early as 1833, finally ceased.

Today almost every mile of track of the Wear Valley Network has been taken up, with the notable exceptions of the stretch from Bishop Auckland to Stanhope, part of an 18.4 mile line to Eastgate in Weardale which serves a cement works there, and the 3 mile stretch of the Heritage Line from Bishop Auckland to Shildon.

THE DARLINGTON AND BARNARD CASTLE RAILWAY

Yet another nominally independent line, the Darlington and Barnard Castle Railway (the D&BCR), was authorised by an Act of 3 July 1854. (It had been projected as early as the 1830s when Joseph Pease undertook surveys of a possible route at his own expense.) In fact, the D&BCR was the first major branch from the S&DR's main line. It ran from a junction at Hopetown, 550 yards north-west of North Road Station, for 15½ miles via Piercebridge and Gainford to a terminus just to the north of Galgate in Barnard Castle. The first sod was cut on 20 July 1854 and the line was opened two years later on 8 July 1856. The line was worked from the beginning by the S&DR, and the two companies amalgamated on 23 July 1858. It is significant that the Enabling Act of 3 July 1854 includes a complete list of the seventeen subscribers, nine of whom hailed from Darlington and eight from Barnard Castle – a discreet balance. No less than three of the Darlington party were members of the Pease family, whose residences were also named – Henry at Pierremont and John and Joseph Whitwell at Southend. Other names intimately involved in the affairs of the S&DR at this time were Alfred Kitching and Thomas McNay.

The continuing power of the landowners was acknowledged in the provision that 'one of the Directors [shall be] appointed by the Duke of Cleveland'. This was a sop to Henry Vane, the 2nd Duke of Cleveland, whose seat at Raby Castle stood in the way of the most direct route. In consequence, he vigorously opposed the project, declaring that it was 'the device of a scheming and Artful Individual [Joseph Pease] to deceive the people of Barnard Castle for his own benefit'. An account of the Duke's opposition at another stage in these protracted negotiations in 1844 is revealing for the light it sheds on the arguments deployed by the landowners on the one hand and the railway promoters on the other when they came into conflict: The Duke 'receives a deputation of persons anxious to bring a railway to Barnard Castle, crossing his Raby estate':

> He replied that he could not see the necessity for these branch lines; it was well enough for through lines, they might be desirable for the country, but branch lines were uncalled for. If a place was within twenty miles of a railway it was all that could be

wished or desired. His Grace also added that when he came to Raby he looked at the beautiful valley of the Tees, and said to himself, surely they will never think of bringing one of those horrid railways through this Paradise of a country. That he did not think a line was wanted for the interest of the country, and would oppose it to the utmost of his power. Mr. [Joseph] Pease reminded him that it was all very well for him, he could order out his carriage at any moment and go with ease and comfort any distance, but every one could not do this, and that the facilities and comfort of a railway were of the utmost importance to the great mass of the population.

The Duke was at ease in his possessions and remained firm in his refusal, and the inhabitants had to submit to the disappointment.[19]

The Duke's opposition led to the first Bill being rejected by the House of Commons in June 1853. However, when the line was opened in July 1856, he attended the customary ceremony, during which he magnanimously expressed the hope that all past difficulties between his family and the S&DR interests would be forgotten. His change of heart may well have been due to the award of a 'directorship for life'!

By a S&DR Act of 16 July 1855 a spur was constructed at the Darlington end, from west to north, Charity Junction to Stooperdale Junction, to enable traffic to work north from the D&BCR onto the S&DR, and vice versa. The Stooperdale Curve was opened on 8 August 1861. The line is wholly in County Durham except for a short stretch just west of Gainford where it unexpectedly crosses over the River Tees to the Yorkshire side. This costly diversion, which involved the building of two bridges over the river within the space of half a mile, resulted from the fierce opposition of the local landowner, the same 2nd Duke of Cleveland whom we have just encountered. This time, the Duke objected to the proposed route because it would bring the line close to his property at Selaby Hall and, more to the point, because of the disturbance which, he maintained, the railway would cause to his fox coverts!

THE STAINMORE ROUTE

The D&BCR was the catalyst for a new and vastly more ambitious scheme – the South Durham and Lancashire Union Railway (the SD&LUR). This project had two objectives. The first, in one direction, westwards, was to transport hard-structured coke from the South Durham coalfield to the ironworks of Millom and Barrow in coal-barren Furness in Lancashire. The second objective had to do with the needs of the Teesside ironworks. As Cleveland ironstone with its high silica content was not of the highest quality, it had to be mixed with richer haematite ores from Ulverstone in South Cumberland, transported by way of a circuitous route to the south via Normanton and Leeds. Similarly, the ironworks at Consett required a proportion of haematite ore, which had to be transported by an equally circuitous route to the north, via the Newcastle and Carlisle Railway and a branch of the NER up the Derwent Valley. The SD&LUR would provide a far more direct route by utilising the existing D&BCR and the S&DR main line to Teesside, and the Bishop Auckland and Weardale Railway to Consett. As a result, the SD&LUR line would be guaranteed substantial traffic in both directions.

Although the line was sponsored by the S&DR, the chief promoter was Henry Pease, in collaboration with several other long-standing S&DR proprietors. The line was estimated to cost £375,000 to construct and the Pease family alone subscribed £15,000 to the project. It was built and opened in stages, and required two Acts of Parliament. In its final form, it ran for an unprecedented 54.3 miles from a junction with the Bishop Auckland and Weardale Railway, at Bishop Auckland, via Barnard Castle and Kirkby Stephen, across the high moorland reaches of the Pennine range, to a junction at Tebay with the Lancaster and Carlisle Railway (the L&CR, worked by the London and North Western Railway (the L&NWR) as part of the west coast route). From Tebay, the SD&LUR had a working arrangement with the L&CR to run over their line south for 16 miles to a junction with the Furness Railway at Hincaster Junction. A similar arrangement was entered into with the Furness Railway for the final 29.2 miles of the route to Barrow-in-Furness (with a 15.6 mile extension to Millom).

The line from Spring Gardens Junction to Tebay was authorised by an Act of 17 July 1857. In the event, it was the Pennine or Stainmore section that was built first, from a junction with the D&BCR 700 yards north of Barnard Castle Station, to Tebay, a distance of 38 miles. The line was engineered by Thomas Bouch, whose brother William was locomotive superintendent of the S&DR, and the first sod was cut by the Duke of Cleveland on 25 August 1857.[20] It was opened for mineral traffic in two stages: as far as Barras (16 miles) on 26 March 1861, and the remainder (22 miles from Barras to Tebay) on 4 July 1861. On that day, six trains hauling 600 tons of Durham coal and coke left the Auckland coalfield for Tebay, while 150 tons of haematite ore reached Teesside from the west via the D&BCR.[21] This Pennine section was deemed sufficiently important to have a formal opening on 7 August, and it was opened for public passenger traffic the next day. The S&DR purchased two 0–4–0 locomotives from Robert Stephenson and Company for working the line –

SIR THOMAS BOUCH 1822–80

Apprenticed as a Civil Engineer between the years 1839 and 1844, Thomas Bouch began a long career as railway designer and builder with the Lancaster and Carlisle Railway from 1844 to 1848. In all, he was responsible for the construction of 300 miles of railways in the north of England and Scotland. In 1849 he was appointed resident engineer of the S&DR. (Subsequently, his younger brother William was appointed locomotive superintendent and engineer of the same company.) Thomas Bouch supervised the construction of the South Durham and Lancashire Union Railway for the S&DR, opened in July 1861 – 'one of the triumphs of mid-Victorian railway engineering'. He was especially proficient in the design and construction of bridges. He pioneered the lattice-girder system of bridge-building, which was deployed to such spectacular effect in the Belah and Deepdale viaducts on the SD&LUR. His plan for floating railways to span the estuaries of the Forth and Tay seem in restrospect to have been auguries of his last notorious commission – the building of the ill-fated Tay Bridge. At 2 miles long and with eighty-five spans, this wrought-iron lattice-girder bridge was the largest in the world when it was opened on 31 May 1878. For this achievement Thomas Bouch was fêted throughout the land, received the Freedom of the City of Dundee, and was knighted. Tragically, the following year, during a violent storm on the night of 28 December 1879, the central portion of the bridge fell into the waters beneath, carrying with it an entire train. Every one of its seventy passengers was killed. Under the shock of this disaster, Thomas Bouch's health gave way and he died on 30 October 1880 in his home town, Moffat, in the Scottish Borders.

No. 160 *Brougham* and No. 161 *Lowther*. They were to a new and more powerful design by William Bouch and Robert Stephenson, to cope with the severe gradients encountered, and were the first bogie tender engines to be used in Britain.

The route chosen for the Pennine section presented formidable physical obstacles. The first was the valley of the River Tees itself, just beyond Barnard Castle. The existing line from Darlington turned from east–west to north–south to reach its terminus facing Galgate, so that an 'end-on' extension was ruled out. Instead, at what became Barnard Castle West Junction, the SD&LUR continued the east–west alignment of a short stretch of the D&BCR. The result was that both railways used that short stretch, from Barnard Castle East Junction to Barnard Castle West Junction. About 260 yards beyond the West Junction, the SD&LUR built its own station so that for a while Barnard Castle enjoyed the blessing of two stations belonging to different 'independent' companies.

Almost a mile beyond the station, the line reached the high east bank of the River Tees, which was spanned by a lofty viaduct, the first of nine to distinguish this route. In their day, they were among the most remarkable feats of railway engineering. Four of the viaducts were constructed using the newly developed lattice-girder system: the Tees, Smardale, Deepdale (well named) and, most spectacular of all, the highest viaduct in Britain in respect of altitude, and the second tallest, Belah, between Barras and Kirkby Stephen. (The tallest was Crumlin, built in 1857 across the Ebbw Valley in South Wales – at 200 ft a mere 4 ft higher than Belah.) The other five stone viaducts were at Percy Beck, Mouse Gill, Aitygill, Merrygill and Pod Gill. The three longest viaducts – Tees, Deepdale and Belah – were built of wrought-iron girders from ironworks on Teesside, for the sake of speedy construction.[22] They were designed by Thomas Bouch, engineer to the line, who also designed the ill-fated Tay Bridge, which was opened on 1 June 1878 and for which he received a knighthood on 27 June 1879 – 'six months and a day before the bridge collapsed'.

It was during the building of lines across the Pennines that the behaviour of the navvies, and the terror they inspired in communities near their shanty towns, was at its worst. The Chief Constable of Westmorland suggested that the S&DR pay the wages of four constables to keep the peace and prevent poaching during the building of the SD&LUR. The company demurred, but grudgingly agreed to pay for one constable at the rate of £1 1s 0d per week plus 1s 6d per month 'boot allowance' and 1s 0d per month for lamp oil. In contrast, the navvies themselves received up to £1 a week and the masons £1 10s 0d, thus the masons received bigger wage packets than the constables who policed them.

From Bowes the line began its long ascent and crossing of Stainmore by way of the valley of the River Greta to the summit at 1,374 ft. The outlook here is bleak in the extreme. In winter, train crews were halted and snowbound for days at a time, on what must have appeared to be the deserted roof of the world. Joseph Pease was not guilty of hyperbole when he commented to his brother Henry, who was destined to become vice-chairman of the company, that if 'the busy, bustling, whistling railway ever traverse[s] Stainmoor's wintry wastes, or the industrialists beyond be supplied with cheapened and excellent fuel, they that profit thereby, and rejoice therein, will doubtless have much to thank thee for in thy exertions and perseverance'.[23]

The Stainmore section from Barnard Castle to Tebay was initially built as a single track, but although the company was far-sighted enough to purchase land sufficient for eventual doubling, perversely three of the viaducts between Belah and Kirkby Stephen were built wide enough for only a single line. (Conversely, Smardale viaduct between Kirkby Stephen and Tebay was built to accommodate a dual track but never did so.) The first working timetable for the 'S&DR – South Durham and Barnard Castle Sections', from 1 September 1861, detailed fourteen Up trains (Darlington to Tebay), four of which terminated at Barnard Castle, and thirteen Down trains (Tebay to Darlington), three of which commenced at Barnard Castle. The trains were numbered so that regulation footnotes could indicate which trains would cross each other at the passing loops at Summit, Bowes and Barnard Castle on this single-track line. Trains were also divided into four categories distinguished by a letter at the head of the appropriate column – M for Mineral, M & G for Mineral and Goods, G for Goods, and P for Passenger.

In contrast to the Pennine section, construction of the Durham section between Bishop Auckland and Barnard Castle was relatively straightforward, although it was formulated in rather piecemeal fashion. The first 1.4 mile stretch between Bishop Auckland and a junction with the Tunnel branch at Fieldon Bridge was authorised by the Second Act of 23 July 1858 and built by the S&DR. The route then made use of the Tunnel branch for a further 1.4 miles as far as West Auckland, and of the Haggerleases branch from West Auckland for 1.2 miles to Spring Gardens Junction. From that point, the line gained height to cross the River Gaunless on a three-arch skewed-brick viaduct that leaped across the valley. One of the arches accommodated the Haggerleases branch, which followed the north bank of the river on the level far below.[24] The bridge itself has been removed, but the imposing brick abutments still remain, together with the lofty circular twin-piers on the north bank. The corresponding pair facing them to the south have collapsed, scattering masses of brick on the grassy slopes.

After 12.3 miles of new construction from Spring Gardens Junction, the line joined the D&BCR at Barnard Castle East Junction, making use of that line for 200 yards as far as Barnard Castle West Junction. This was a minor role in the fortunes of the SD&LUR which we can credit to the D&BCR. A major role was to provide an outlet east via Darlington for the Pennine section of the SD&LUR while the Durham section was being built. It also carried iron ore from Ulverston in Lancashire to the ironworks on Teesside in its own right. The newly built Durham sections, and consequently the whole of the route from Bishop Auckland to Tebay, were opened for mineral traffic on 1 February 1863 and for passengers on 1 August 1863. In the meantime, the SD&LUR had been amalgamated with the S&DR (not unexpectedly) on 1 January 1863, by an Act of 30 June 1862.

The building of the section of line between Bishop Auckland and Fieldon Bridge Junction provided a direct rail connection between Bishop Auckland and West Auckland. It was then no longer necessary to work trains between these two towns along the Bishop Auckland and Weardale Railway to Shildon Tunnel North Junction, reversing and then proceeding along the Tunnel branch. After almost seven years, passenger services between Shildon Tunnel North Junction and Fieldon Bridge Junction ceased.

Mineral traffic, of course, continued, some of which was strategically important to the SD&LUR. There were coke ovens attached to Adelaide Colliery, with access to the line near Shildon Tunnel North Junction, and at St Helen's Colliery, with access near West Auckland. We recall that the transport of coal and *coke* to the blast furnaces at Barrow and Millom was a prime consideration in the building of the SD&LUR.

Penrith to Workington

With the development of blast furnaces at Workington on the Cumberland coast 36 miles north of Millom, a new chapter began in the history of the SD&LUR. The incentive to reach Workington was identical to that for reaching as far as Barrow and Millom in the first place – to capitalise on the transport of hard-structured Durham coke westwards to the ironworks there, and haematite iron ore from mines at Whitehaven eastwards. It was no easy task. The distance between Cockermouth (the nearest railhead) and the most convenient junction with the SD&LUR, at Kirkby Stephen, even as the crow flies was 39 miles. The physical, geographical obstacles in the way were, if anything, even greater than those that challenged the builders of the Stainmore route to Tebay. The whole of the Lake District as well as outliers of the Pennine range intervened. In the event, two separate railways were authorised to bring the project to a successful conclusion – the Eden Valley Railway (the EVR) and the Cockermouth, Keswick and Penrith Railway (the CK&PR).

The nominally independent EVR was promoted by the S&DR and authorised by an Act of 21 May 1858. It ran from a junction with the SD&LUR at Kirkby Stephen for 24 miles to a junction with the L&CR 3½ miles south of Penrith. (From 10 September 1859 the L&CR was leased to the L&NWR.) The first sod was cut at Kirkby Stephen on 4 August 1858. Francis Mewburn noted in his *Larchfield Diary* that at the dinner given at Appleby the day before, 'The memory of Edward Pease was drunk in solemn silence. No such honour was ever given to a Quaker since the days of George Fox.' The line was opened for mineral traffic on 8 April 1862, and for passengers on 9 June 1862. At the west end of the line, two curves were built to connect with the L&NWR. As the CK&PR, with which the EVR was intended to connect, was not completed, a south-facing junction with the main line at Clifton was built first. (At the insistence of the L&CR, 'nervous that the seemingly innocuous nature of the Eden Valley line might conceal grand designs for a new trunk route seeking to reach Scotland'.)[25] Nevertheless, Clifton Junction enabled traffic from the EVR to travel south to Tebay on L&CR (L&NWR) metals, and vice-versa, providing a welcome alternative to the direct route from Kirkby Stephen, and reducing bottlenecks on these single-track lines.

Just west of Appleby the line crossed the Midland Railway's main line – the renowned Settle and Carlisle – and a short east to south spur was constructed to enable traffic to transfer from one line to the other. The EVR was single track throughout except for this Appleby spur and the triangular junction with the L&NWR.

The CK&PR was authorised on 1 August 1861. It was opened for mineral traffic on 26 October 1864, and for passengers on 2 January 1865, that is, after the S&DR was amalgamated with the NER on 13 July 1863. It was a 'splendidly scenic line' of 31¼ miles, engineered by Thomas Bouch, who was responsible for the SD&LUR, and the only one to cross the Lake District. It ran from Penrith via Troutbeck, the Greta Gorge, Keswick and Bassenthwaite to Cockermouth, Wordsworth's birthplace.[26] When the CK&PR was completed, a loop line was built from Weatherrils Junction on the EVR line to the north-facing Clifton North Junction (later Eden Valley Junction) on the L&NWR main line. When this was opened on 1 August 1863, it completed the Clifton triangle and enabled traffic to travel north from the EVR and then over L&NWR metals for 1.6 miles to Eamont Bridge Junction. (The original intention was to continue into Penrith parallel to the main line, but this was dropped when the L&NWR agreed to grant running powers.) From Eamont Bridge an Act of 23 June 1864 authorised the NER to build the mile-long Redhills Curve to connect with the CK&PR at Redhills Junction.[27] It was brought into use on 5 September 1865, and so completed the through route from Kirkby Stephen to Cockermouth and avoided trains having to reverse at Penrith.

The CK&PR shared Cockermouth Station with the L&NWR, successors to the 8¾ mile Cockermouth and Workington Railway, which was authorised on 21 July 1845 and which provided the last link in the chain of the far-flung route, from West Auckland to Workington. By an Act of 30 June 1862, both the South Durham and Lancashire Union Railway and the EVR together with the Frosterley and Stanhope Railway, were amalgamated with the S&DR on 1 January 1863.

When the CK&PR was completed, it was possible to travel across England by rail 'coast to coast', from Workington in the west to Saltburn in the east, for 122 miles, through some of the most beautiful countryside in the United Kingdom. The dramatic scenery of the Lake District traversed by the CK&PR led on to the marginally less scenic but tranquil Eden Valley, which lent its name to the second railway along the route. At Kirkby Stephen the SD&LUR took over to convey passengers to Barnard Castle, over the wild and windswept Pennine watershed by way of Stainmore, thence, conveyed by the D&BCR, down Teesdale, especially favoured by the Romantic poets and writers, not least Sir Walter Scott, and finally to the cliffs and beaches of the Yorkshire coast at Saltburn-by-the-Sea on the S&DR, whose promoters, in stages, and in one way or another, masterminded the entire route. Shortly after the first Barnard Castle Station became redundant with the building of the second, its fine classical portico was removed, renamed the Albert Temple, and re-erected by the Saltburn Improvement Company at the furthest end of the S&DR, in Valley Gardens, Saltburn. It is still there, a symbol and a reminder of the vision and the enterprise of our railway forefathers who forged an iron road from sea to sea in the course of a single generation.

Just as a route 'from Consett to the Cumbrian Coast' was brought about by the completion of the Stainmore section, so the completion of the Wear Valley Complex had earlier established a route 'from Consett to the North Sea Coast' at Redcar, which was to have unforeseen consequences. As W.W. Tomlinson remarked:

> When the Derwent Iron Company, soon after the opening of the Cleveland ironstone district, took a lease of the Earl of Zetland's royalty of Upleatham, the S&DR Company had reason to congratulate themselves on the quite fortuitous circumstance that the system at one point touched the very edge of the ironstone field, and at another, 54 miles away, was in contact with the principal ironworks in the County of Durham. That the ironworks at one end of the system should draw their supplies from the ironstone field at the other end of the system was about as perfect an arrangement from the S&DR point of view as could be imagined.[28]

THE NORTH YORKSHIRE IRONSTONE LINES

The ironstone field to which Tomlinson referred was situated in North Yorkshire, and in particular in the Cleveland Hills. In the course of time, it was to generate an iron and steel industry close by on the banks of the River Tees which would far outstrip the Consett Complex. Indeed, for many years 'Teesside' was the largest and most productive concentration of iron and steel works in the world. Ironstone had been worked in North Yorkshire for many hundreds of years, from at least the twelfth century. However, it was the discovery of the Cleveland main seam in 1850 that led to the phenomenal growth of the industry on Teesside. That growth in turn stimulated the growth of a network of railway lines to sustain it.

On 8 June 1850 John Vaughan of the firm of Bolckow and Vaughan, together with the company's mining engineer John Marley, was prospecting for ore in the Eston Hills, on the estate of Sir J.H. Lowther. In a quarry used for roadstone, only 6 miles from Middlesbrough, the two men came across a solid rock of bare ironstone some 16 ft thick – the Main Seam. It was John Vaughan's appreciation of the immense significance of this discovery of the 'Great Cleveland Ore Field' which entirely transformed the situation. At once, an expansion of iron-ore mining was put in hand. In 1851 the Eston Hills mines produced no less than 188,000 tons of ore. The main seam was then progressively worked eastwards in a belt through Upleatham, and beyond the Eston Hills to Skelton and Loftus. As the richer seams were exploited, mining tended to migrate southwards to less productive seams. From 1853 mines were established in a belt along, and at the foot of, the north-facing scarp of the Cleveland Hills, from Whorlton in the west through Ingleby and Roseberry to Guisborough in the east, and thence further east, as the scarp receded through Boosebeck to Skelton, Liverton, Skinningrove and Loftus.

The fortunes of the railways and of the ironstone mining and quarrying industry in Cleveland were therefore very much bound together. From its beginning in 1825, the S&DR had quickly demonstrated an ability to handle large tonnages of mineral traffic, and it is not surprising that it was the first of the 'public' railways to expand into Cleveland on the promise of revenues from the transport of iron ore. Conversely, the chief incentive for prospecting for ironstone on the north-facing side of the Eston Hills was the overriding need for a source close to an established railway line (the S&DR Middlesbrough–Redcar Extension).

Five interdependent industries contributed to the production of those iron goods that furnished the sinews of the Industrial Revolution: the ironworks themselves; mining for ironstone; for coal; for limestone; and not least, the railways. It is no coincidence that a few powerful and successful individuals and companies had major financial interests in some or all of these industries. As an example of this interdependence, coal owners from County

HENRY WILLIAM FERDINAND BOLCKOW 1806–78

The iron and steel industry on South Teesside has historically been synonymous with the names Bolckow and Vaughan. Henry William Ferdinand Bolckow was born at Sulten, Mecklenberg, North Germany, in 1806. When he was twenty-one, he moved to England where he subsequently became a naturalised subject. At first he was employed by the firm of C. Allhusen, corn merchants, of Newcastle upon Tyne, which had been founded by two brothers, friends of Henry Bolckow, who had emigrated from Rostock near Mecklenburg. Bolckow left Newcastle to go into business on his own and settled in Middlesbrough in 1841. There he entered into an illustrious and long-standing partnership with John Vaughan (1799–1868). The two men met when they began courting two sisters, whom they duly married, and the two families lived next door to each other from 1841 to 1860. Bolckow and Vaughan complemented each other. Bolckow was the businessman and entrepreneur, in which respect he had much in common with Edward Pease, and Vaughan, the older man, while he had little financial knowledge was the practical scientist and engineer. He had a wealth of experience producing iron in South Wales, the Black Country and in Newcastle, where he ultimately became manager at Losh, Wilson and Bell's works at Walker on Tyne. After Middlesbrough acquired its charter of incorporation in 1853, Bolckow became its first mayor. He was unanimously elected Middlesbrough's first MP in November 1868. He was very generous to his adopted town – in 1867 he presented 72 acres of his estates in Middlesbrough for a public park named in honour of Prince Albert. Bolckow built a mansion (since demolished) at Marton on the outskirts of Middlesbrough as his family home, surrounded by an estate that is now the municipal Stewart Park. Henry Bolckow died at Marton in 1878.

Durham, including the Quaker families of Darlington, were among the first to work the stratified iron ores of Cleveland, where similar mining skills and expertise were called for.

The Pease family's investment in both coal mining and the S&DR had marched hand in hand and prudently served each other. The same considerations came into play when, through the agency of Pease and Partners, the family expanded their activities to include ironstone mining in Cleveland. Holding a commanding position in railway transport when the worth of the Cleveland main seam ironstone was recognised in the early 1850s, they bought or leased mines around Guisborough, from 1857 in the Eston Hills, and from 1865 in the Skinningrove area to the east.

It is significant that Darlington was the home and headquarters of Pease and Partners and of the S&DR – two important contributors to the success of the iron industry on Teesside – as well as the conduit for a large proportion of its raw materials. Its place in the history of what was described between 1830 and 1880 as 'the largest iron foundry in the world' is assured.

The transportation of raw materials to, and of finished products from, the ironworks played a large part in determining their location. The railways were essential for the carriage of coal, ironstone and limestone, and sites close to railways were therefore critical. To a large extent this mutual dependency affected the financial outlook of the two industries in good times and bad.

The railways responded to the challenge presented by the expansion of the iron industry on Teesside with the building of three strategic lines, and the adaptation of a fourth existing line, for ironstone traffic, each tapping ever more distant sources of iron ore. Two of these – the Teesside Lines and the Middlesbrough and Guisborough Railway (the M&GR) – were part of the S&DR domain and are discussed below.

The Teesside Lines

The Middlesbrough and Redcar Railway (the M&RR), opened on 4 June 1846, was already fortuitously in existence, running parallel to the northern edge of the Eston Hills and only some 2½ miles distant. Shortly after the discovery of iron ore there, Bolckow and Vaughan themselves built a private temporary tramway, followed by a permanent branch line, from the junction of three inclines at the foot of the escarpment, north to Eston Junction (Grangetown) on the M&RR. As the result of strenuous efforts, the Eston Mines line to the Cleveland Ironworks and Grangetown Station was opened for traffic as early as December 1850. By 1858 this private railway had been extended under the Middlesbrough to Redcar tracks and onto a jetty built at the side of the main channel of the river, to provide an alternative outlet for the ore and an inlet for seaborne supplies. As in the case of the coal industry in south Durham, the Eston Mines line was the first of a number of private systems built in situations where either the mine owners could not wait for the 'public' railway companies to extend their tracks, or where there seemed little justification for the expense involved in serving a small mine by a public system.

The Derwent Iron Company had the lease of a very productive ironstone working at Upleatham on the north-east slopes of the Eston Hills. When the mine opened in 1851, the company built a winding narrow-gauge railway 5 miles long to a junction with the M&RR at Coatham, half a mile to the west of Redcar Station. When the Derwent Iron Company ran into financial difficulties in 1857, J.W. Pease and Company (later Pease and Partners) took over the lease. Then, when the Redcar and Saltburn Railway was built in 1861, the S&DR constructed a mile-long standard-gauge branch, which replaced the narrow-gauge railway, from 'banktop' at Upleatham to a junction just to the west of Marske Station.

The Middlesbrough and Guisborough Railway

The nominally independent M&GR was the first of three major ironstone lines in North Yorkshire. It was initially promoted to exploit mines in the Guisborough area to the south of the Eston Hills. Although the aged Edward Pease noted in his diary in September 1851 the recently announced proposal for 'a railway near Guisborough',[29] the S&DR management committee was cautious and rejected the whole thing as 'chimerical'.[30] Fortunately, that view was not held by Pease's son Joseph and Joseph's eldest son, Joseph Whitwell (1828–1903), who personally promoted the line as a demonstration of the family's own faith in the potential traffic revenues to be derived from ironstone shipments. The line was authorised by an Act of 17 June 1852 and ran for 10 miles from Guisborough Junction, 550 yards east of Middlesbrough Station, by way of Ormesby, Nunthorpe, Pinchinthorpe and Hutton Gate to a terminus in the centre of Guisborough town. At the head of the company's first share certificates a vignette depicts Guisborough Priory at one end of the line and blast furnaces at the other. In the foreground is an 0–6–0. locomotive hauling a train of iron-ore wagons, with the hill called Roseberry Topping in the background.[31]

At first, money for the building of the line was scarce and Joseph and his son were obliged to step in with a contribution of £5,000. This funding shortfall was only remedied, however,

---

SIR JOSEPH WHITWELL PEASE 1828–1903

A grandson of Edward Pease, 'father of the railways', and the eldest son of Joseph Pease, Joseph Whitwell took over the leading role in the affairs of the Pease family when his father relinquished control.[32] Together they promoted a number of railways, including the M&GR – in the face of much opposition from members of the S&DR board. Joseph Whitwell had a rather special, personal connection with this enterprise. He was one of the few members of the family who chose to live outside Darlington, building a mansion at Hutton Lowcross near the M&GR line, 1½ miles from the terminus in Guisborough. Joseph Whitwell married Mary Fox of Falmouth on 23 August 1854, shortly after the line was opened, and it was probably after he had brought his bride to live at Hutton Lowcross that, as a principal shareholder, he had a private station built at Hutton Gate to serve his mansion there. The drive from the hall and part of the station buildings are now part of a private house. After Joseph Whitwell's death Hutton Gate became a public station and remained so until the line closed in 1964.

Lithograph portrait of Sir Joseph Whitwell Pease (1820–1903).

when the Peases offered a guaranteed dividend. Joseph Pease cut the first sod on 3 October 1852 and the line was opened for mineral traffic on 11 November 1853, and for passengers on 25 February 1854. From the beginning, the line was a great success, especially since mineral traffic boomed. The M&GR had only a short life as an independent company as it was amalgamated with the S&DR, which had worked the line from the beginning, on 23 July 1858. (No doubt the S&DR management committee overcame its initial reservations as soon as the line was demonstrated to be a commercial success.)

Members of the Pease family continued to be heavily involved in the affairs of the M&GR. Edward and his sons Joseph, John and Henry were principal shareholders, as was Joseph's son, Joseph Whitwell. Joseph Whitwell and Mary Pease had eight children. The eldest son, Alfred Edward, born in 1857, later settled at Pinchinthorpe House, just over a mile to the west of his father's home at Hutton Lowcross. In Alfred Edward's edition of his great-grandfather's *Diaries* he wrote: 'There is now [1906] at Pinchinthorpe Station a wooden step, that was made for Edward Pease to get in and out of the train when he travelled to and from Ayton.' Edward Pease was a member, and for a while Chairman, of the Committee of Management of Ayton Agricultural School, in whose affairs he took a keen interest. Pinchinthorpe was then the nearest station to Great Ayton.

The M&GR Act also authorised the building of three short branch lines in the Guisborough area to serve ironstone mines in the Cleveland Hills escarpment there. One of them, to serve mines near Roseberry Topping, was never built. A second diverged 700 yards west of Hutton Gate and ran for 2 miles via Hutton Village and then east to Cod Hill mine. A short spur led off westwards to the Hutton mines. This branch was opened at the same time as the main line, on 11 November 1853. Passing as close as it did to Joseph Whitwell Pease's mansion at Hutton Lowcross, this branch must have been a constant, visible reminder of his investment in the surrounding ironstone industry. A third 1.2 mile narrow-gauge branch brought iron ore down from Belmont mine via Belman Bank directly to Guisborough goods station yard. Much of this iron ore was destined to travel on over S&DR metals as far afield as the blast furnaces at Tow Law in County Durham.

CHAPTER 9
Rival Railways

The Great North of England Railway

Eleven years were to elapse from the opening of the S&DR before another railway penetrated the Darlington area, hitherto the preserve of the S&DR. As early as December 1826 the *York Herald* published a proposal that the S&DR should be extended from Croft Bridge to the City of York.[1] The eventual outcome was the GNER Company's projected line from York to Newcastle upon Tyne, which was an important stage in the realisation of George Stephenson's dream of an east coast route from London to Newcastle and the Borders. It was to be 'a continuation of all the proposed lines from the metropolis towards Scotland'. It might seem strange that from the beginning the Pease family and their Quaker partners in Darlington were prime movers in the promotion of the GNER, since it constituted the first invasion of the S&DR's territory. Nevertheless, according to W.W. Tomlinson, the GNER project, which was announced to the investing public in October 1835, was the product of 'the deeply-planning brain of Joseph Pease'.[2]

There were two reasons why the project was especially attractive to the proprietors of the S&DR. First, 'it would secure for the Company a valuable strategic position in the developing trunk rail network of eastern England'.[3] Second, it would increase the company's mineral traffic by means of 'a cheap and expeditious transport of coals into the heart of the North Riding and to the City of York itself; as well as . . . a communication between the port of Stockton and every part of the Kingdom'.[4] It is noteworthy that following the announcement of the project every S&DR shareholder received a copy of the GNER's prospectus, together with a statement from the managing committee 'that it is exceedingly desirable that those who are interested in the S&DR should also become shareholders in the said undertaking . . . and thus [bring] in aid of the direction of the new Concern . . . that experience which may eventually promote the interests of both Companies'.[5] The GNER has therefore been described, with good reason, as 'the child of the S&DR'. Indeed, the proposal so incensed that thorn-in-the-flesh of the S&DR, Christopher Tennant, that he was provoked to declare that the company 'were to get the lion's share of the benefit from the GNER'.[6]

The line was authorised in two stages. First – because it was realised that this would be the more difficult section to construct – the northern section from Newcastle through Darlington to Croft, by an Act of 4 July 1836; and second, the southern section from Croft to York, by an Act of 12 July 1837. Joseph Pease and his fellow promoters sought the advice of Robert Stephenson, who urged them to concentrate their resources on completing the easier and cheaper southern section in order that the company could begin to earn revenue, and to abandon work on the northern section. The promoters took his advice – even to the extent of selling back to the original owners land that had been acquired for the railway north of Darlington. When the promoters announced in August 1837 that the originally intended order of construction was to be reversed, the County Durham shareholders strongly objected, fearing a number of ulterior motives on the part of the S&DR interests, including a desire to protect their Yorkshire markets for Auckland coal. Notwithstanding, the first sod was cut near Croft, on the southern section, on 25 November 1837.

In order to facilitate the original intention to build the northern section first, the whole of the Croft branch of the S&DR had been bought for £20,000 in 1839. In the event, only the 1.4 mile section from what is now Croft Junction through Darlington to Albert Hill (Parkgate Junction) came under GNER control. (The remaining 2.1 miles to the terminus at Croft reverted to S&DR control. Together with running rights over the section utilised by the GNER this enabled the company to maintain its branch line service.) This transaction enabled the southern section of the GNER from York to Croft, plus the 5 mile stretch from Croft to Darlington, to be opened for mineral traffic in both directions on 1 January 1841. When a passenger service commenced on 30 March 1841, Darlington was at last linked by rail to London, and George Stephenson's dream was a stage nearer realisation. (The passenger service was extended to North Road Station on the S&DR line, with a shuttle service making use of the Croft branch tracks as far as Parkgate Junction.) The first trains were hauled by four S&DR locomotives, two on the northbound and two on the southbound trains. The pilot engine for the Darlington–York run, appropriately, was an S&DR locomotive, No. 26 *Pilot*, a new 0-6-0 Timothy Hackworth-designed locomotive manufactured by Kitchings. Other S&DR locomotives worked on the GNER, either on loan or on hire.

By early 1841, the GNER had used up all its capital and had reached only as far as Darlington from York. All in all, the editor of the *Railway Times* was justified in writing that 'the GNER . . . appears to have been, in the outset, one of the worst-managed undertakings in the Kingdom, and that is saying a great deal'.[7]

With the northern section in limbo it is not surprising that a new company, the Newcastle and Darlington Junction Railway (the N&DJR), had to be incorporated on 18 June 1842. The *éminence grise* behind this proposal was George Hudson, the 'Railway King' (1800–71), whose empire centred on York, and he was duly appointed Chairman of the new company.[8] Technically, the GNER in its northern section incarnation was still in existence. Discredited by its decision to abandon that section and build the southern first, and the much publicised resignation of Thomas Storey as engineer to the project, the company was more than willing to negotiate a leasing arrangement with George Hudson's N&DJR. Not unnaturally, Joseph Pease and his colleagues on the S&DR were outraged when they learned that Hudson had usurped the power they had expected to wield north of Darlington in pursuit of his overweening ambitions. Consequently, they bitterly opposed the N&DJR proposals, which their representative, Captain Watts, denounced as 'an abortion with a crooked back and a crooked snout, conceived in cupidity and begotten in fraud'.

With Robert Stephenson appointed as engineer-in-chief, the new company promptly took over the building of stretches of line needed to complete the northern section of the GNER's route from Newcastle to Darlington, an arrangement ratified by an Act of 11 April 1843. In order to make use of portions of three existing lines the route was a circuitous one.

In the event, when this section from Darlington northwards was opened on 18 June 1844,[9] it stopped short of Newcastle on the southern bank of the River Tyne at Gateshead. Nevertheless, it was the realisation of the dream shared by George Stephenson and George Hudson to link the Thames with the Tyne. 'Nothing, next to religion,' opined one of the company's directors, 'is of such importance as a ready communication.' With a typical Hudson flourish, the 'Grand Opening Train' on that historic day set out from London and reached Gateshead at an average speed of 37 mph. This achievement was notwithstanding the fact that the route, entirely over the metals of companies in which George Hudson now had a controlling interest, was much more circuitous than the present direct east coast main line. This trunk route was the 'jewel in George Hudson's crown'. At the height of

> ## GEORGE HUDSON 1800–71
>
> The 'Railway King' was born on 10 March 1800 in the village of Howsham on the River Derwent in the East Riding of Yorkshire. The son of a tenant farmer, at fourteen years of age George left home and was apprenticed to a draper in York. Shortly before his twenty-first birthday this uneducated but hard-working young man was made a partner in the firm, and that same year he married the proprietor's sister, Elizabeth Nicholson. George seemed content with his role as silk mercer, but in 1827 he was left a small fortune by his great-uncle Matthew which transformed his life and launched him on an unprecedented entrepreneurial career. Hudson's first involvement with railways occurred in 1833 when he was elected Treasurer of a company to bring a railway line into York, later abandoned in favour of a proposed York and North Midland Railway (the Y&NMR) with Hudson as Chairman. (He once famously boasted that he 'would make all the railways come to York'.) 'He was on the threshold of one of the most amazing rises to personal power and wealth in British industrial history.' It was Hudson's good fortune that he came on the scene when there was no such thing as a national plan for railway development. He was shrewdly able to take full advantage of the opportunities for speculation and profit which the railways afforded, and which in turn precipitated the 'Railway Mania' of 1844–7, with all its disastrous consequences. Conscious of his working-class background, Hudson aspired to and achieved recognition by the high society of the day. He was elected the first Conservative Lord Mayor of York in 1837 and MP for Sunderland in 1845. He was host to Queen Victoria and Prince Albert, and became a member of the Carlton Club. He was consulted on business affairs by the nobility, including the Duke of Wellington, and bought Londesbrough Hall near Market Weighton from the Duke of Devonshire as his country seat, complete with private railway station. Notwithstanding, Hudson's fall from grace was as inexorable as his rise to fame. Accused of fraud, he was arrested and imprisoned for debt. For seventeen years, from 1854, he was obliged to spend long periods in exile abroad to escape his creditors. When he finally returned to England in 1870 it was as a poor man. He died the following year, and his body lies 'in a humdrum grave, in an obscure provincial plot' in the churchyard of Scrayingham close by Howsham, the place of his birth.

his powers, he controlled 1,450 route miles of line out of the total of 4,000 miles then built.[10]

The N&DJR crossed the S&DR at Albert Hill, half a mile south-east of North Road Station. During the lifetime of the S&DR and its successors, including the L&NER, this was one of the few right-angle crossings on the level on United Kingdom main lines.[11] The first NER Rule Book, inherited from the practice of the S&DR, gave very explicit instructions regarding the working of this crossing, including the injunction that 'the coal and mineral trains are invariably to give way to the passage of passenger trains'. Every driver had to 'sound his whistle at least half a mile before reaching the crossing, and continue to do so until he has the attention of the signal-man'. Speed over the crossing was then limited to 10 mph. Any infringement of the rules was to be reported and 'the signal-man to be liable to a fine of Five Shillings in every case in which he shall omit to report any engine driver or other person not complying with these rules'.

S&DR Crossing, Albert Hill, Darlington, photographed in 1925.

Originally, there were spurs in all four angles of the crossing, but only the south-west one remains, rather larger than the original because it is further out from the actual crossing. It joins the north–south line at Parkgate Junction. The GNER engine shed was located inside the south-east spur, and a rebuilding of 1854 can still be seen by the side of the east coast electrified line, but in a sadly derelict condition. All else, including two separate signal-boxes and redundant track, has been swept away.

The first station on the new line was a mean wooden shed erected by the GNER on a site in Darlington which later became the present Bank Top Station of 1887. When Queen Victoria stopped here on 28 September 1849 she was presented with a loyal address by Francis Mewburn, on behalf of the citizens of Darlington. This did not prevent her from criticising the appearance of the station 'in the place which gave birth to the railway system'. A few improvements had been made by the time the Queen stopped again on 28 August 1851, when she was presented with a bouquet of flowers, and again on 30 August 1852, when she received a basket of fruit.

The GNER had its offices in an elegant Georgian building on the south side of the junction of Crown Street and Northgate, opposite those of the S&DR on the north side. There is evidence in a painting of the King's Head Inn by William Dresser that it also occupied premises next door to the inn on the south side, where Northgate becomes Prebend Row. They are identified in the painting by a sign that reads 'Great North of England Newcastle and Darlington [*sic*] Railway Company's Parcels Office'.

Trunk Route Competition

We have seen that when the GNER was first proposed the S&DR did not see this as an invasion of its territory. On the contrary, the former was largely seen as the child of the latter. Indeed, as a *quid pro quo* for the ceding of the Croft branch, the S&DR obtained running rights over GNER metals from Parkgate Junction to Croft Junction through Bank Top Station, which was a much more convenient site in relation to Darlington town centre than its own station at North Road – as indeed it remains to this day.

The resurrection of the northern section of the GNER in the form of the N&DJR was an altogether different proposition, as it enlarged the empire of the S&DR's arch rival, George Hudson, and at the company's own expense. The bitter opposition of Joseph Pease and his colleagues took practical effect in the promotion of a rival route to the north shortly before the N&DJR was incorporated on 18 June 1842. On 22 April 1842 a poster appeared advertising a 'Railway Communication between London, Darlington and Newcastle' from 5 May. That heading, however, was misleading and disingenuous, as such a communication could not be entirely by rail. Although the proposed communication was to begin at the GNER's station at Darlington, it was promoted by the S&DR. Hence the route 'will be by the S&DR to South Church, near Bishop Auckland'. This claim was only made possible by the opening of the Shildon Tunnel Railway to South Church a few days earlier, on 19 April 1842.

From South Church the journey north was to be 'by well-appointed [horse-drawn] omnibuses'. The remains of the platform at South Church, where passengers transferred from omnibus to train and vice versa, can still be seen by the side of the track just to the south of the railway bridge over the A188. Passengers were conveyed by omnibus the 14½ miles (as the crow flies) to Rainton Meadows, north-east of Durham City, the temporary terminus of the Durham Junction Railway, which communicated with two others to form a route as far as Gateshead-on-Tyne. As all three companies were nominally independent at the time, the S&DR was free to enter into an arrangement with them fulfilling the poster's promise to 'communicate' with Newcastle. Ironically, these were the very companies featured in George Hudson's plans for a route to the north, and were subsequently acquired by the N&DJR.

Notwithstanding the purple prose of the poster, this was a highly inconvenient and circuitous route from Darlington to the Tyne. Not surprisingly, it was unsuccessful and soon abandoned. One suspects that it was never envisaged as a viable, practical proposition and only justified as stealing a march on the S&DR's powerful rival. (In order to attract customers, the S&DR even put a man on the GNER's Bank Top Station to meet passengers from the south and try to persuade them to continue their journey by the S&DR route.) When the N&DJR's route from Darlington to Gateshead was finally opened on 18 June 1844 it had become clear that not only would the two companies have to coexist, but that they would also have to cooperate with one another. The provision of the spurs at the S&DR Crossing enabled passengers and goods traffic to transfer from one system to the other without a break. It was at this time that the S&DR Crossing took its name. The controlling signal-box, located hard by the track-side in the south-east angle, followed suit with 'S&DR Crossing' on its name-board. On the poster, the service to Newcastle was advertised as 'in connection with Trains to and from Dinsdale Baths [*sic*], Yarm, Stockton and Middlesbrough'. The connection was made at North Road Station, from which the shuttle service ran to Bank Top Station via Parkgate Junction.

THE LEEDS NORTHERN RAILWAY

The second invasion of S&DR territory took place 8 miles to the east and a decade later. It was the crossing of the River Tees by the great 42-arch brick-and-stone viaduct designed by Thomas Grainger, at Yarm, in 1849, which brought the LNR into S&DR territory. High as the viaduct was, it was not high enough to bring the track level with the top of the bank on the Durham side of the river. Almost immediately the line was obliged to tunnel under the Yarm Coal Depot branch just short of its terminus, before ascending gradually to encounter the S&DR main line on the level near Whitley Springs, Egglescliffe. By 1852, the S&DR had no doubt learned to be reconciled with rivals as powerful as the LNR. The intruder successfully negotiated permission from the incumbent to cross its line at this point. As a *quid pro quo* the LNR built two extra lines alongside its own for 2 miles north to Mount Pleasant where the two lines diverged. The LNR leased these parallel tracks to the S&DR for 1,000 years at a nominal 1*s* per annum. The S&DR took possession on 25 January 1853, and the stretch of the original S&DR main line through Preston Park thus becoming redundant, was abandoned.[12]

When the crossing at Egglescliffe was completed, regulations similar to those controlling the S&DR Crossing in Darlington were issued. Passenger trains took precedence over goods and minerals, and a speed limit of 10 mph was imposed. The regulations were signed by Thomas McNay for the S&DR and by Samuel Smiles for the LNR. The Crossing was later replaced by a conventional junction, appropriately named Preston Junction, when the LNR was dissolved by the NER Act of 31 July 1854 and vested in the York, Newcastle & Berwick Railway (the YN&BR, once part of George Hudson's empire), which then became a constituent of the NER. Since the Yarm Coal Depot branch and the LNR trunk route were on different levels, there could be no connection between them to enable a transfer of traffic between the S&DR and the LNR. To remedy the situation, Allen's Curve was constructed, on land purchased from Leonard Raisbeck in 1853, in the south-west angle of Preston Junction. This enabled trains from Darlington to travel south over the LNR line and vice versa. Some would stop at the LNR's Yarm Station, located immediately to the south of an over-bridge that carried a by-road to Aislaby and Middleton St George. This station also was therefore located not in Yarm (despite its name) on the Yorkshire side of the river, but in Egglescliffe on the Durham side. In S&DR and NER days, Allen's Curve was largely used by iron-ore trains originating on the North Yorkshire & Cleveland Railway and destined for ironworks in County Durham.[13]

The LNR built North Stockton Station in Bishopton Lane (site of the present Stockton Station), so called to distinguish it from *South* Stockton Station (now called Thornaby) on the Middlesbrough Extension line. To avoid north-bound traffic from Middlesbrough going all the way round by Preston Junction, the S&DR constructed a curve from Bowesfield West Junction to the LNR line at Hartburn Junction. Although it was authorised by an Act of 13 August 1859, it was not opened until 1 May 1863.

THE CLARENCE RAILWAY

The Clarence Railway was an anachronism from the beginning. It started life as early as 1824 under the grandiloquent title of the Tees and Weardale Railway, which expressed the Stockton party's aspiration for a direct line to the Auckland coalfield. Whereas the S&DR was initially conceived as a localised line to serve the internal coal trade, in its final form the Clarence Railway set its sights on the lucrative potential of an export trade in coal from the

Tees. To further this objective it also set its sights on tapping the West Durham coalfield further to the north by means of branches from its main line north-west towards the City of Durham. The change of name, shortly before the company's incorporation, was designed to honour the Duke of Clarence, later King William IV.

As projected, the main line was to run from a junction with the S&DR at Sim Pasture Farm (2 miles east of New Shildon), for 14 miles via Stillington to terminate at Haverton Hill on the River Tees, at a point that the company named Port Clarence (or more popularly and historically 'Samphire Batts'), opposite Port Darlington. Ironically, Haverton Hill was one of the two options considered by the S&DR as the terminus for a Middlesbrough Extension Railway, but rejected in favour of Port Darlington. The Clarence Railway was the brainchild of Christopher Tennant, who rightly saw the S&DR's proposed extension to Middlesbrough as a threat to the borough of Stockton. Its avowed object was 'to open a shorter course than heretofore between several valuable Mines of Coal and the River Tees'.[14] Tennant produced a persuasive table of comparative distances which highlighted the advantages of the proposed Clarence Railway over the S&DR route.

The spirited opposition of the S&DR to the Clarence Railway was somewhat mitigated when the latter received the support of Henry Blanshard, a shareholder in the S&DR and a prosperous and influential London merchant. The Clarence Railway Act duly received Royal Assent on 23 May 1828, less than three years after the opening of the S&DR. The line was opened piecemeal for private owners in stages as it was built. Although the usually accepted dates fall between August 1833 and January 1835, in fact, the first coal for export reached Stockton on 29 October 1833, Haverton Hill in January 1834, and the final terminus at Port Clarence, with its newly constructed coal staithes, towards the end of 1834.[15]

The Clarence Railway was the only west-to-east rival to the S&DR's main line, and it was a good deal shorter. As laid down, from Simpasture Junction to Stockton the route was only 12 miles compared with the S&DR's 18.3. As it threatened the S&DR's monopoly of the lucrative coal trade from south-west Durham, virtually by way of the direct route proposed by the 'Stockton Party' and so vigorously opposed by the 'Darlington Party', the S&DR in turn did all in its power to obstruct the new line. First, and within a month of the opening, it prohibited the horse leaders of the Clarence Railway from operating over its lines during the hours of darkness, from one hour after sunset to one hour before sunrise, although its own leaders were not prohibited.[16] Second, its own loaded wagons were just counted, while those of its rival were stopped and subjected to delay, having to be weighed and checked at the Thickley weigh-house before leaving the S&DR line. Finally, the Clarence Railway horse leaders were obliged to adhere strictly to the regulations while a blind eye was turned to horse leaders on the S&DR line.[17]

Despite these obstacles, the Clarence Railway began to make heavy inroads into S&DR traffic. Its Achilles heel, however, was that it had no independent access to the pits, and it was obliged to make use of the S&DR's line as far as Simpasture Junction. Accordingly, the S&DR imposed on the colliery owners a levy of two-pence farthing the ton-mile on coal passing over its lines and then destined for the 'Clarence route' to Stockton. Not surprisingly, the owners chose the cheaper S&DR route, and the Clarence Railway lost its trade and was greatly discomfited and financially embarrassed, to the extent that the directors were obliged to borrow £100,000 from the Exchequer Loan Board to complete the construction of the line. Nevertheless, it struggled on, at first utilising horse power and from 1838 steam power, mainly for the passenger service.

During its lifetime, the Clarence Railway put out a number of branches. One of the most important was the so-called City of Durham branch, which diverged at Stillington West

Junction and terminated at Ferryhill. This was later extended to Byers Green, where it made an end-to-end junction with the West Durham Railway (the WDR).

The fortunes of the Clarence Railway and the WDR were very much bound up with each other, as the WDR provided the first part, and the Clarence Railway the second part, of a strategic outlet from the lucrative and productive south-west Durham coalfield (centred on Crook) to Port Clarence near the mouth of the River Tees. Once again this brought the Clarence Railway in direct conflict with the S&DR, which had hitherto enjoyed a monopoly of the south-west Durham coal trade through its subsidiary, the Bishop Auckland and Weardale Railway. It therefore looked askance on this interloper into its territory, as did the Pease family, who had extensive mining interests in the Crook area. The S&DR consequently did everything in its power to obstruct the joint venture. When construction on the WDR was well advanced, it was obliged to promote a retrospective West Durham Railway Act, passed on 4 July 1839, which made the acquisition of land, in the face of the S&DR's opposition, much easier.

Despite set-backs in its early years, it could be said that the Clarence Railway, in the guise of the West Hartlepool Harbour and Railway Company, with which it amalgamated on 17 May 1853, outlasted its rival the S&DR, for while the former was absorbed by the NER in 1865, the latter had been absorbed two years earlier in 1863.

THE CLEVELAND RAILWAY

The third large ironworks on Teesside, after Bolckow and Vaughan and Dorman Long, was that of the Bell Brothers at Port Clarence on the north bank of the River Tees, built on land owned by the Stockton and Hartlepool Railway. The company title was derived from three brothers, Isaac Lowthian (later Sir Isaac), John and Thomas, whose father Isaac Wilson was a partner in the firm of Losh, Wilson and Bell, ironmasters at Walker-on-Tyne. While Thomas seems to have attended chiefly to the company's coal-mining interests, and Isaac to have been the chemical metallurgist of the ironworks, John was the principal figure in their ironstone mining activities in Cleveland, where they owned or leased mines at Normanby, Skelton and Skinningrove.[18] The owners were obliged to send ore from the Normanby mine southwards by way of a 1.4 mile narrow-gauge line to a junction with the M&GR near Morton Grange, and thence by way of Middlesbrough and Stockton to ironworks at Port Clarence on the north bank of the River Tees. Their objective, therefore, was to promote a railway that would bring iron ore from the eastern mines all the way to Middlesbrough via Guisborough and Normanby (incidentally replacing the narrow-gauge line), as an alternative to the problematic route by sea. The proprietors could have chosen to build a line to the S&DR's railhead at Redcar, to the north of the Eston Hills,[19] or to the M&GR's railhead at Guisborough to the south. The drawback in either case was that they would be handing over revenues for the second, substantial stage of the journey to the S&DR or its subsidiary.

It may seem strange that the prime mover behind the projected railway was not the Bell Brothers themselves, but Ralph Ward Jackson, who took over the mantle of Christopher Tennant as the 'inveterate opponent of the S&DR'. He had set his sights on breaking the S&DR's monopoly in Cleveland, thus inaugurating the 'Struggle for the Cleveland Ironstone District'.[20] Ward Jackson and the Bell Brothers together decided to build a direct line of their own all the way from Skinningrove to Middlesbrough by way of Guisborough. The second part of this route, of course, brought them into direct conflict with the S&DR, the parent company of the rival M&GR. Not surprisingly, the Durham and Cleveland Union Railway

Bill of 1857, as it was first called, which provided for trains crossing the Tees on a steam ferry, was hotly contested in Parliament by the S&DR, and was thrown out by the Lords. Some idea of the antagonism between the two parties can be gauged from a letter from the normally conciliatory John Pease to Ralph Ward Jackson:

> Whilst I admit that the commerce of the kingdom is open to all and that *a fair and honourable competition* is not to be thoughtlessly deprecated, it would seem to me impossible that thy policy could be regarded as such . . . thy movements have ever seemed to be a series of agressions at once unprovoked and unreasoning. The Stockton and Darlington Railway has been most unblushingly assailed and in a way which they conceive has no parallel in the history of railway conflict; not one single stone would appear ever to have been unturned or opportunity missed to diminish the Stockton and Darlington company's resources by abstraction of traffic. They are not aware on the other hand that upon any occasion they have intruded into your district or acted otherwise than strictly on the defensive.[21]

When the promoters finally received authorisation in an Act of 23 July 1858 – after 'an intense parliamentary struggle fought with consummate skill and self-possession by Ward Jackson and Joseph Pease'[22] – it was a compromise, since it covered only the first section from Skinningrove to Guisborough. As Tomlinson remarked, it was something of a drawn battle.[23]

In the meantime, the proprietors decided to press ahead with the Guisborough to Middlesbrough section, so as not to be at the mercy of the S&DR, relying on way leaves for which no Parliamentary sanction was required. The situation *was* regularised, however, with the passing of the Cleveland Railway Extension Act of 27 July 1861,[24] by which time the extension, engineered by John Hawkshaw, was almost completed.

The Cleveland Railway Extension as authorised was 8¼ miles long and commenced near Belmont Farm, just to the south of Guisborough. After crossing the M&GR line 570 yards south of its terminus at Guisborough Station, it curved and ran directly west to Upsall Grange alongside the main Middlesbrough Road, parallel with the rival line from Guisborough to Nunthorpe. (The remains of earthworks belonging to both lines can still be seen on either side of the A171.) From Upsall Grange, the line turned sharply to run northwards over the western flank of the Eston Hills at Flatts Lane and Normanby. Having crossed the Middlesbrough to Redcar main line it terminated at a jetty on the Normanby Estate at Cargo Fleet, from which loaded wagons of ore were transported over the Tees in open barges.[25] The line was opened for mineral traffic on 23 November 1961. A spur was built in the south-west angle of the crossing to make a connection with the Middlesbrough to Redcar line.

There is a story, no doubt apocryphal, that supporters of the S&DR threatened to blow up the rival Cleveland Railway's bridge over its line in Guisborough when it first came into use.[26] There was certainly intense rivalry between the two companies since the Pease interests had invested in the M&GR in the expectation of hauling large quantities of ironstone all the way from East Cleveland to ironworks as far distant as Tow Law in County Durham. However, that rivalry did not stand in the way of the M&GR pragmatically permitting a short connecting spur to be built from its line to the Cleveland Railway, skirting Guisborough to the south. It extended for 480 yards from Hutton Junction on the M&GR in the west to Belmont Junction, at the commencement of the Cleveland Railway Extension, in the east. No doubt the proprietors of the M&GR saw merit in a connection, as it would permit wagonloads of iron ore from mines in East Cleveland to be diverted at Guisborough from the

Cleveland Railway to their line, for destinations on the S&DR network near and far. If they initially sanctioned the connection in the hope that the Cleveland Railway Extension itself would never be built, they would be disappointed, since the extension and the connection were both opened on the same day – 23 November 1861.

A glance at the map indicates that, apart from the incursions of the four rival railways dealt with above, the S&DR's empire remained remarkably compact and undisturbed. It was with some truth that a biographer of George Hudson remarked: 'By 1846 there was barely a mile of line north of Leeds and east of the Pennines that was not Hudson's, *except for the stubbornly independent Stockton and Darlington Railway*.'[27]

AMALGAMATION: THE END OF THE LINE?

'Stubbornly independent' as the S&DR remained, it was not immune from the pressures of contemporary railway politics and was bound sooner or later to succumb to a railway rival more powerful than itself. The S&DR only narrowly escaped losing its independence comparatively early in its career. Surprisingly, the initiative came from the S&DR itself: driven by a crisis in its financial affairs, at some stage early in 1849 the management committee invited George Hudson to negotiate for a lease of its lines. By May of that year, a Bill for the lease of the S&DR by the YN&BR, of which Hudson was Chairman, was actually prepared. Fortuitously, as it turned out, it failed to come to fruition on account of George Hudson's fall from grace, at which point the management committee of the S&DR prudently withdrew its participation in the leasing bill. It had been touch and go!

The next suitor for the S&DR was a much more formidable proposition. The NER was formed on 31 July 1854 by the amalgamation of the Y&NMR, the YN&BR and the LNR. At a stroke, it became the first of the great regional railway monopolies, controlling 720 route miles extending through much of North Yorkshire, Durham and Northumberland. 'It possessed the most extensive, yet geographically compact railway network in Britain.'[28] The NER already had various points of contact with the S&DR, and sought an amalgamation – largely in order to thwart the L&NWR's aim of establishing routes to the east coast ports.[29] Moreover, the east–west configuration of the S&DR network was an ideal complement to the north–south orientation of the NER's own lines.[30] Relations between the two companies had always been cordial, neither provoking the other by attempted incursions into their territory. Indeed, it was Henry Pease's proud boast in 1863 that the S&DR 'had never made a yard of railway in a district already accommodated by railways'.[31] An amalgamation would suit the interests of both parties, so that the hitherto independently minded S&DR viewed the NER's approach with favour, and negotiations began in December 1859 at the York headquarters of the NER. The S&DR delegation was led by John Pease and had as its main objective the establishment of full local control over the old S&DR network. By March 1860, agreement had been reached in principle.[32] The two companies began to work together from 1 January 1861, and an amalgamation was formally agreed at a meeting of the S&DR board on 13 February 1863.[33] It was duly authorised, after a smooth Parliamentary passage, by an Act of 13 July 1863. 'Thus, after a life of 42 years, for 38 of which it had been operating trains, the S&DR ceased to exist as a separate entity.' Since the S&DR was the 'First in the World' it is regrettable that it was not powerful enough to absorb other rival railways in the north-east, but had itself to be taken over by a company thirty years its junior. By way of compensation, the NER and its successor the L&NER were justly proud of this part of their common ancestry. When significant anniversaries of the S&DR's inaugural day recurred the opportunity was seized to celebrate that beginning with a suitable sense of history and in some style, as we shall see.

At the time of the amalgamation the S&DR and its subsidiary companies had built up a network of 201½ route miles. When we add a further 54.8 miles of running rights over lines such as those of the Furness Railway we arrive at a total of 256.3 operating miles. Lines the S&DR was authorised to build but which were not completed at the time of the amalgamation added a further 31.7 miles – a grand total of 288 miles. This was the substantial dowry brought by the S&DR to add to the NER's existing network. (By the time of the amalgamation the company had abandoned or sold 11.8 miles of line – a total built, authorised, or enjoying running rights from 1825 to 1863 just one-fifth of a mile short of 300.)[34]

In the event, the amalgamation made little practical difference to the running of the S&DR's domain, since the Act provided that these lines should be operated by a *committee* composed in the main of former S&DR directors, but including two from the NER. Conversely, three S&DR members of the committee also had seats on the board of the NER. The Darlington Committee was granted almost complete autonomy within its own area. Its meetings were held at Darlington North Road Station, and here the committee 'adjudicated on all the various interests connected with the line'. It had the power to appoint its own Secretary, Solicitor and all officers and servants, and to fix the times and number of trains, fares and rates. In 1863 Henry Pease was appointed Chairman of the Darlington Committee, an office he held until its demise.[35]

Although expected to last for a limited period of ten years, with a possible extension of a further two years, the Darlington Committee remained in being until 1876 and its end appears to have been brought about by the death of William Bouch, one of the remaining stalwarts of early S&DR days. Certainly, the committee ceased to function shortly after his death, although it lingered on, with a nominal existence only, until 1879. During the lifetime of the committee, the area under its jurisdiction was known as the Darlington Section and on maps of that period lines continued to be marked 'North Eastern Railway: Darlington Section'. Eventually, it became known as the Central Division of the NER. The King is Dead, Long Live the King!

CHAPTER 10

Darlington – Firstborn of Railway Towns: a Portrait

INDUSTRY, HOUSING AND AMENITIES

The motto that appeared on Darlington's coat of arms, *Floreat Industria* – 'Let Industry Flourish' – was particularly apt.[1] One industry that certainly flourished was the railway industry, so much so that we are entitled to ask, to what extent can Darlington be characterised as *the* railway town?

In his essay 'The Concept of a Railway Town and the Growth of Darlington', B. Barber defined the criterion by which a railway town might be identified as follows: 'That its origin and growth was determined by and dependent upon the employment potential created by the establishment of railway company works for the manufacture of capital equipment.'[2] According to this strict definition Darlington was *not* a railway town in the same league as say Crewe, Swindon and Wolferton. These towns were created *de novo* by railway companies on 'greenfield sites'. From the beginning, therefore, they were obliged to provide necessary urban facilities for their employees, especially housing but also churches, schools and leisure services. The same was true, but to a lesser extent, of such railway towns as Ashford, Eastleigh, Horwich and Earlestown.

By contrast, in 1825 Darlington was already an important, prosperous, and well-established town of some 7,200 inhabitants, with a history of settlement and expansion going back to early medieval times. It was not summoned into existence under the 'enchantment of the rod which Stephenson and Pease – the Moses and Aaron of the new age – wielded and waved'.[3] As Barber points out, in demographic terms Darlington in 1801 was already a far larger settlement than any of those that were to develop as 'classic' railway towns, and throughout the following century it maintained this superiority, although Crewe and Swindon were its equal in size by 1901.[4]

In 1825 Darlington's prosperity depended on three factors: its position on a main arterial highway – the Great North Road; its place as a marketing centre serving a large rural, agricultural hinterland; and its textile industries. In particular, the local linen industry, which had flourished in the preceding century and which by 1790 was believed to be larger than that of any town in England and Wales, made it a centre of regional importance. The fact that the population of Darlington grew relatively rapidly in the 1820s and '30s was almost certainly due to the parallel growth of linen and woollen manufacture. These were precisely those industries upon which the wealth of the Pease family and their Quaker business associates was also based.

When those same Quakers were instrumental in bringing the railway to their home town in 1825, therefore, it was in no sense dependent on the employment that this infant industry brought in its wake. The locomotive works of the S&DR was first located in Shildon, and in any case for much of its history the company relied upon outside contractors for the majority of its engines. Nevertheless, two local Darlington firms, W. & A. Kitching and William

Lister, did expand to undertake the building of locomotives and related repair and maintenance. It is to them that we must look for the first effects of the coming of the railways in the field of employment, apart from the relatively small number of men directly employed by the S&DR. The physical invasion of Darlington by the railway system, including the GNER in 1841, had produced no dramatic effect on its economic organisation by 1851. The reason was that Darlington still possessed no major railway works by this date, in contrast to other railway towns, and notwithstanding the fact that it was the centre of operations for the two rival companies, the S&DR and the GNER. In terms of employment, census returns in 1851 and 1861 reveal that the two companies together had only 154 and 266 identifiable employees living in the town (that is to say 3 and 4 per cent respectively of the working population), although this does not take account of the number of labourers who would be employed at the companies' warehouses and goods yards.[5]

If we put on one side Barber's rather strict definition of a railway town and apply more general criteria, it is evident that by 1851 Darlington was becoming, if it was not already, the kind of railway town that a visitor of the times would instantly recognise. The most visible and immediate sign was the provision of houses for railwaymen and their families in two new communities – at Hopetown near North Road Station in the case of the S&DR, and near Bank Top Station in the case of the GNER's successor, the YN&BR. Their lines had intentionally been built at some distance from the existing township. Thus John Dobbin's celebrated painting of the opening of the S&DR in 1825, in the rural setting of Skerne Bridge, has as its counterpart a picture by the same artist of the meadows between Bank Top Station and St Cuthbert's parish church. In time, the two new communities became suburban nuclei developed round the two stations and their goods yards. In 1849 the S&DR suburb at Hopetown had a population of 260 and the corresponding GNER suburb at Bank Top had a population of 1,185. If we assume, with some justification, that the two districts were largely inhabited by railway employees and their dependents, then in 1849 14 per cent of the total population of the borough was maintained by the two railway companies.[6] By that parameter, Darlington had already come of age as a railway town.

However, it is important to note that the essential housing for railwaymen and their families was not provided by the railway companies themselves, as an act of altruistic patronage. The distinctive railway terraces were built in the main by speculators for rent, and therefore for profit. In 1849 the S&DR owned only nineteen cottages and one public house. It may also seem surprising that the company did not initially provide the kind of amenities as, for example, the GWR did for its employees at Swindon. The reason is not far to seek. The Quaker faith that animated the Pease family and their friends and associates in Darlington traditionally involved a concern for social welfare, and the physical, spiritual and mental well-being of the community as a whole, and not merely their own employees. Its practical expression was the provision of schools for the children, leisure facilities for the men, and opportunities for part-time employment for the women. Thus the Darlington Mechanics Institute – opened in 1853 and one of the first of its kind in the country – was largely funded by the Pease family. Its stately classical facade still fronts Skinnergate opposite the Quaker Meeting House. It had a hall that could seat 500 people. Similarly, the Darlington Free Public Library in Crown Street was built as the result of a legacy of £10,000 for that purpose in the will of Henry Pease of Pierremont. The foundation stone was laid in June 1884 by Joseph Whitwell Pease, and the building was opened by Henry Pease's daughter Henrietta in October of the following year.

As the S&DR was dubbed the 'Quaker Line', just so Darlington was a Quaker town with public amenities – including in due course the Town Hall – paid for by influential Friends.

Nevertheless, the S&DR did establish a Railway Institute with imposing premises on the junction of Whessoe Street and North Road adjacent to the North Road Shops. By the end of the century, it possessed an impressive library of 3,000 books, which was no doubt a valuable supplement to the resources of the Mechanics Institute and the Public Library.

For its part, the moral responsibilities felt by the GNER directors resulted in an appeal to their shareholders for a building fund for a church to serve the railway community around Bank Top. The site chosen was a prominent one close to the station, in the angle of Neasham Road and Yarm Road. The foundation stone was laid on 10 September 1847 by George Hudson, Chairman of the GNER's successor the YN&BR. During the Railway King's later tribulations the distinguished William Etty of York testified in his defence: 'I knew him to be kind-hearted, generous and public-spirited to the last degree. I was only twice in my life in a Railway carriage with him. And on one of these occasions . . . Mr Hudson gave £400 or £500 to a Church at Darlington, for the Railway men, without hesitation.'[7]

This church was designed by the architect John Middleton (1820–85) in the dignified Early English style of Gothic church architecture, with a dominant west tower. It was opened on 3 January 1850 and dedicated to St John the Evangelist. As William Etty suggested, the church was built principally for the convenience of the railway workers and their families, that is 'for the accommodation of a poor population', who lived in the terraced rows near the station. Unfortunately, they did not appreciate the church's peal of six bells, especially if the men had been on nightshift. The church is still there, in good order, at the centre of the parish of East Darlington.

If we now revert to housing as the criterion for a railway town, the one event that put Darlington on the map in that respect was the opening of the North Road Shops for the building and repair of locomotives on 1 January 1863, at the very end of the independent career of the S&DR. Although only 150 men were initially transferred from Shildon, by 1865, under NER auspices, the number of employees had risen to 391 and to about 1,000 a decade or so later. In its heyday, the works became Darlington's largest employer with 3,500 men. This rapid expansion of the workforce led to a corresponding development of standard terraces to house them and their families in the immediate vicinity of the Shops. In due course, this new development, to the north of the main line, merged with the pioneering railway cottages of Hopetown to the south, until this railway enclave, like that at Bank Top, was in its turn swallowed by the built-up expansion of Darlington to boundaries far beyond its medieval core.

The names of the many remaining streets around Hopetown betray their origins: Station Road, McNay Street, Stephenson Street, Shildon Street, Gurney Street and Foundry Street. Those at Bank Top express a patriotic attachment to the royal family, which flourished at the time they were built: Victoria Street, Albert Street, Adelaide Street, King William Street and Prince's Street.

When we extend this study to include the history of the NER, and its successor the L&NER, from the absorption of the S&DR by the former in 1863 to the nationalisation of the latter in 1947, it is apparent that Darlington earned a place among the first rank of railway towns. (The tide of railway expansion only began to ebb after the Second World War, and then only gradually.) During this period of some eighty-five years, the visible signs of the railway's presence and influence grew in number and in scale. This was especially true of the two wide corridors of industry along the railway routes which crossed each other at Albert Hill. Here the extensive workshops, the large depots and the widespread marshalling yards of the railway companies were joined by many railway-related industries as well as independent heavy industries such as iron foundries and rolling mills. They were attracted by the numerous economic benefits the railways bestowed. Not least among them, of course, was the ability to transport their raw materials and finished products to and from all parts of the kingdom along lines radiating out from a railway centre that had achieved national, and not merely regional, importance. Strictly speaking, Darlington as a railway *centre*, a nodal point, began in 1844 when the GNER, metamorphosed into the N&DJR, was built across the S&DR at Albert Hill. It was Darlington's good fortune that this first, south to north incursion into S&DR territory was later to become part of the east coast main line trunk route, in preference to the parallel line of the LNR through Stockton – not for the first time did Stockton lose out to its larger neighbour.

The unprecedented growth of Darlington in the 1860s was thus achieved through a major structural change in industrial emphasis, brought about by the coming of the S&DR, in which the hegemony of iron replaced the primacy of the textile industries. As the author of the 1871 census observed: 'The increase of population in the Darlington district [from 11,033 in 1841 to 27,729 in 1871] is not due to an increase in the old manufactories of Darlington, viz. cotton and woollen mills, etc., but to the introduction of the manufacture of iron, to the erection of blast furnaces, rolling mills, forge works, engine building works and other small manufactories.'[8] Of course, conversely, the S&DR and its successors were themselves the beneficiaries, as industrial expansion in the region increased the volume of goods traffic by rail, accelerated by the incentive to expand the railway network much further afield.

Although the period from 1863 to 1947 lies beyond the independent history of the S&DR, it is legitimate to select two typical developments, in 1887 and 1901 respectively, that underline Darlington's emergence as a premier railway town in the second half of the nineteenth century and the first half of the twentieth.

Bank Top Station, 1887

On its formation in 1854 the NER was in a much stronger position than its predecessors to do something about the mean condition of Bank Top, the most important station on its premier line. Nevertheless, thirty-three years were to pass before the company more than made up for its neglect. The new Bank Top was designed by the company's own architect, William Bell, and opened on 1 July 1887. No expense was spared, down to the smallest detail of the fittings and furnishings, and a very handsome and worthy station it proved to be. Fortunately, after well over a century of continuous use, it remains largely unaltered to this day.

Overall, there are three graceful glass and iron-framed train-shed roofs. The facade of the main building is in red brick with stone quoins at the angles. The main entrance is on the west side facing down Victoria Road, which was built to connect the town centre with its new amenity. No doubt this wide approach road was named, and the station was opened, in 1887 in honour of Queen Victoria's Golden Jubilee, which was celebrated that year.

The grand entrance consists of two immensely high and wide archways, designed to permit the passage of the carriages of the wealthiest of first-class passengers, with their coachmen sitting aloft. These arches opened into a spacious vestibule or *porche-cortere*, which, at four bays deep, was large enough to enable carriages to enter by one arch, swing round in a semicircle, and depart by the other. The vestibule is protected from the weather by a train-shed-style roof, in imitation of those above the station proper. On either side of the main entrance arches, to left and right, are two smaller arches for the convenience of foot passengers. In between the arches are the lower stages of a soaring, Italianate clock-tower, 90 ft high, which is quite literally Bank Top's crowning glory. It can be seen from every part of the town. The lower stage consists of two storeys with two windows one above the other facing east; the

Bank Top Station, Darlington, designed by William Bell and opened on 1 July 1887.

second stage has windows to all four quarters; and the third stage has clock faces to north, south, east and west. All the windows have galleries beneath them, as have the clock faces, which also have triangular pediments above. The final stage is a pyramidal roof – almost a spire – with a round window in each gable and surmounted by a gilded flèche. Historically, the clock was set to run five minutes fast to ensure that hurrying, anxious passengers always caught their train. A large bell, cast by John Warne and Son of London in 1886, was installed to toll the hours. (One suspects that it was instigated by the architect William Bell as a punning reminder of the part he played in this enterprise.) For many years, the bell was disused and in store until, in 1983, it was placed on a plinth in public view on platform four.

From the vestibule a subway leads beneath the main Down line to emerge onto an immensely long island platform, which at 640 yards – one-third of a mile – is one of the longest in Britain. Each span of the triple-arched train-shed roof is 60 ft wide and 1,000 ft long. They curve slightly at their north end and cover, respectively, the Down main line on the west, the island platform with the principal station offices in the centre, and the Up main line. The train-shed roofs are supported on parallel rows of eighty massive cast-iron Corinthian columns. As an indication of the company's attention to detail, there are three cast-iron painted and gilded shields of arms arranged in a clover leaf pattern in each spandrel of the arches. The spandrel to the left of each column contains a shield of arms of Darlington Borough at the top, Stockton Borough to the right, and Durham County to the left. The spandrel to the right of each column similarly contains three shields in a clover leaf pattern which together constitute the arms of the NER, viz. the shields of arms of the City of York in the centre, the LNR to the left, and the YN&BR to the right. There are 158 spandrels in total, and since each cast-iron shield has a representation of the arms on the front and the back, that makes for a total of no less than 948 coats of arms.

After a dispute with porters, more passengers began to use what was originally the rear entrance to the station, which faces an approach road from Parkgate to the north, between the Up and Down main lines, terminating at the north end of the island platform.

Prior to 1887, the operation of trains to Stockton from the first Bank Top Station was highly inconvenient – north to Parkgate Junction, then east along the original S&DR route. With the building of the new station the opportunity was taken to construct a new line, which diverged from the main line 350 yards south of Bank Top at Polam Junction. From there it curved to run east for 3 miles and 410 yards to join the S&DR route at Oaktree Junction. The new line was opened at the same time as the new station, on 1 July 1887. From that date, passenger trains ceased to run on the 3½ mile stretch between Albert Hill and Oaktree Junction.

After sixty-two years, this was the first time that any part of the original S&DR locomotive-hauled route was abandoned to any but mineral traffic. It is ironic that on the day that Darlington came of age as a railway town with the opening of Bank Top Station, a substantial part of the world's first public steam-hauled passenger railway there should be abandoned. The bonus, however, is that it is now possible to walk the whole of this stretch, from the east side of the east coast main line, close by the old GNER locomotive shed, to just short of Oaktree Junction. Walkers who attempt the route will be in no doubt of their starting point: clearly visible on the west side of the tracks of the east coast main line there is a large sign for the benefit of passengers which reads 'Route of the Stockton and Darlington Railway 1825'.

Robert Stephenson and Hawthorn Limited, 1901

The locomotive building works of Robert Stephenson and Company, which began life on 23 June 1823, was principally founded on the prospect of orders for steam engines from the pioneer S&DR. It continued on its Forth Street site in Newcastle upon Tyne for more than three-quarters of a century until, its premises there becoming too cramped and inconvenient to sustain the volume of orders, a historic decision was taken in 1900 to transfer operations to a greenfield site – appropriately on the northern outskirts of Darlington. An entirely new factory was built at Springfield to the east of the L&NER main line, with rail access, between what is now Thompson Street East and the River Skerne. At the same time, a self-contained estate was built for the employees between the works and Salters Lane South. The first locomotive was completed in October 1902, and 'Stivvies', as it was affectionately known, became one of the largest manufacturers of steam locomotives in the world, mainly for export.

Almost as long established in the steam locomotive building business as Robert Stephenson and Company was R. & W. Hawthorne-Leslie and Co. Ltd. The firm was founded in 1817 by Robert Hawthorne, joined later by his brother William, on a site at Forth Banks in Newcastle upon Tyne. R. & W. Hawthorne began as marine and steam engine builders and engineers, but the brothers entered into the locomotive business in 1832 when they delivered their first engine to the S&DR – No. 19 *Darlington* of the Timothy Hackworth-designed 'Director' class.

The removal of Robert Stephenson and Company to Darlington in 1901 gave Hawthorne's the opportunity to extend and improve their own factory, as part of the Stephenson premises were taken over for that purpose. The two firms merged in 1937 and assumed the composite title Robert Stephenson and Hawthorne Ltd, concentrating their main-line steam locomotive building at Darlington and industrial steam locomotive building at Newcastle. In 1955 the company was taken over by English Electric, which continued to manufacture locomotives under the old name. In 1957 the firm switched to the production of diesel-electric locomotives, so ending an unrivalled 154 years of continuous production of steam engines.

Robert Stephenson and Hawthorn out-shopped its last locomotive, D 6898, in April 1964 and the following month both the works and the housing estate were sold. Three years later, on 2 April 1966, the North Road Shops of British Rail also finally closed. For sixty-four years Darlington had uniquely played host to two of the largest locomotive building works in the kingdom. On that ground alone it could fairly claim to be among the first rank of railway towns.

Quality of Life

The standard terraced houses provided for railway employees at Hopetown and Bank Top were of the most primitive kind. Typically, there would be no more than four rooms to each dwelling – 'two up and two down'. Each house would have its own small yard at the rear with the privy, an outside earth closet, next to the coalhouse at the far end. Each had its own hatch to provide access from the back alleys or lanes that separated two rows of houses, back to back. The privy hatch was for the convenience of the night soil collectors, and a second hatch was for coal, delivered from the horse-drawn carts of the merchants whose depots were a feature of every railway station goods yard. The back alleys served many other purposes: the only playground for the children; a meeting place for railwaymen's wives to gossip and talk shop; the place where clothes lines were stretched from wall to wall on washing day – and woe betide the unfortunate coal merchant who chose to deliver on that dedicated day of the week. Alan Suddes, in his commentary on *The Godfrey Edition Old Ordnance Survey Maps: Darlington South*, observes:

> The residents of the immediate vicinity of St John's church were almost all engaged in work associated with the railways. Just to the north of Yarm Road were the engine sheds at the end of Green Street. Many drivers, firemen and cleaners occupied these cramped terrace houses, so handy for the sheds. The tenants of the houses in this area had to get used to continual dirt and grime from the railways and also the noise of passing trains and marshalling and shunting operations which went on through the night.

Away from the immediate vicinity of North Road and Bank Top Stations, similar terraces were built to house the workers of North Road Shops, other railway ancillary industries, and the heavy industries that had been attracted by the presence of the railways, in some cases cheek-by-jowl with the works themselves. From the mid-nineteenth century, terraced housing development was extended down the western slope of Bank Top, especially after the new station was built there in 1887, occupying land from the railway to the River Skerne, from Smithfield Road in the south to Haughton Road in the north. Similar development occurred at Albert Hill and Harrogate Hill, to the north and east of North Road Shops and to the south and west of North Road Station.

The development of working-class housing occurred in the north and east of Darlington, within or adjacent to the industrial corridors. Further out, a band of rather better middle-class houses was developed, occupied by artisans and semi-professional social groups, especially those who were employed by the railway and its associated industries. A good example is an area along Yarm Road once known as the Freeholders' Estate, which was bought in 1851 by John Harris, a Quaker railway engineer, and sold off as individual building plots, each plot having a rateable value of £10. This enabled those of modest means to live in what was to develop as the area comprising Pease, Cobden, Bright, Harris and Milton Streets. The houses here had extensive gardens and provided a striking contrast to the tightly packed streets of Bank Top. 'Many thrifty and long-serving senior railmen established themselves in these family houses. Here we find signalmen, drivers, guards and firemen living alongside a Wesleyan minister and the editor of the *Darlington and Stockton Times*.'

In direct contrast to the back-to-back terraces in close proximity to the works, and even the middle-class houses just described, the residences of the wealthy Quaker businessmen and industrialists, who had provided the impetus and the capital for the railways (from which they, in turn, derived much of their income), were situated in 'the country', as it then was, to the west and south of the town centre. Their large, solid and for the most part plain mansions and villas were built in extensive landscaped grounds with long tree-lined and stone-walled boundaries. Many of the names of these mansions were evocative of the countryside. No less than seven, owned by members of the Pease family, had locomotives named after them: No. 58 *Woodlands*; No. 62 *Southend*; No. 64 *Larchfield*; No. 98 *Pierremont*; No. 118 *Elm Field*; No. 127 *Polam*; and No. 128 *East Mount*. Other Quaker residences included Beechwood, Feethams, Brinkburn, West Lodge, Hollyhurst and Hummersknott.

Later, when the first generation of the Quaker dynasties had passed on and their heirs departed to live elsewhere, these estates were sold for housing development. Fortuitously, most of the houses thus created were provided with gardens, and many green spaces were left, and this part of the town, especially in the neighbourhood of Woodlands Road, Carmel Road and Coniscliffe Road, retains a rural atmosphere to this day. The names of many of the present leafy roads and avenues perpetuate the names of these erstwhile Quaker mansions, while others bear the names of members of the family of the Duke of Cleveland, the one-time landowner, as a condition attached to the sale of the estate to developers. Similarly, the concentration of housing within and adjacent to the railway corridors to the north and east left much of the rural character of Darlington's southern and western approaches unspoilt. 'Thanks to the influential and wealthy Quaker families many of their large 19th century mansions with extensive grounds have now been developed and provide numerous green spaces in a town which has survived the ravages of the Industrial Revolution well.'[9] The railway companies' works and those of many railway-related industries having almost entirely disappeared, it may be that the most abiding visible legacy of that era is this pattern of housing which had its origins in the railway cottages of the S&DR.

One principal element associated with housing and which had a considerable impact on quality of life also had a historically significant connection with the railways. We have seen that a major factor in the creation of the S&DR was the need for an efficient and economic way to transport coal from the collieries of south-west Durham to towns spread out along the River Tees. One of the few branch lines authorised in the Enabling Act of 1821 was in Darlington and it led to a depot, where coal for domestic consumption was stored. Indeed, the first wagonloads were transported on the inaugural day itself and distributed free to the poor of the town. Later, a number of other coal depots were established in the borough, either alongside or connected to railway lines. The *Godfrey Edition* of the OS maps of 1898 gives the location of five such depots. Most of this coal went to fuel the fires of the cheek-by-jowl workers' houses in the industrial corridors, contributing to the pall of smoke that hung continuously over these concentrations of humanity. The pollution of the air from the smoke of domestic fires and factory chimneys was aggravated by the volume of smoke pouring forth from innumerable locomotives hauling freight and passenger trains, or shunting in the marshalling yards. One can understand in retrospect why it was a condition of the Rainhill Trials that the competing engines should be able to 'consume their own smoke', even though the best endeavours of locomotive designers from Timothy Hackworth onwards to achieve that with coal-fired locomotives came to nought. During the era of steam, smoke pollution became an accepted 'occupational hazard' of every railway town. Because of the direction of the prevailing wind, the more affluent areas of Darlington to the south and west were relatively less affected, but when the wind blew from the north and east even they suffered the consequences.

The withdrawal of steam from British Railways in the 1960s, coupled with the passing of the Clean Air Act of 1956 with special reference to house fires, led to a drastic reduction in the consumption of coal. Not only did the quality of the air in towns such as Darlington dramatically improve, but a less obvious consequence was the closure of most of the lineside coal depots that had been such a familiar feature of former days.

It is a paradox that the *demise* of the railway industry in Darlington – which had been the source of so much of its prosperity in times past – should in this way contribute, albeit indirectly, to an improvement in quality of life for its residents.

THE COAT OF ARMS

The coat of arms of the Borough of Darlington as granted on 20 May 1960, and incorporating Locomotion No. 1 *and the town's motto,* Floreat Industria.

The most obvious and widespread symbol of the S&DR is seen wherever Darlington's coat of arms is displayed. When Darlington became a borough in 1867 it devised its own distinctive coat of arms. At the top of the shield was a stylised version of *Locomotion* together with its tender and a chauldron wagon, to symbolise the railway industry. Unfortunately, as was pointed out in the 1920s by Mr Arthur Fox-Davis in a letter to the *Darlington and Stockton Times*, 'The 1867 arms are absolutely bogus and devoid of authority.' On 25 May 1960 a new and legitimate coat of arms was approved by the College of Arms, which retained *Locomotion* and her tender. However, the new coat of arms did not survive for long. With local government reorganisation in 1974 and the resultant boundary changes, the enlarged Borough of Darlington was granted a new coat of arms that reproduced elements of the old County Borough arms and those of the former Darlington Rural District in its design. Once again, *Locomotion* with her tender was retained.

The steam engine *Locomotion* has the distinction of appearing 'in chief', that is, at the top of the shield, and, deservedly, on all three versions of the borough's coat of arms. Residents and visitors alike are thus reminded of the S&DR wherever the shield is displayed, indoors and out. As we have seen, in its oldest version it appears no less than 158 times in the spandrels of the train-shed arches at Bank Top Station alone. It is also reproduced, in a much larger and more readily visible form, at the top of the eight pillars that support the parapets of the twin railway bridges over Parkgate, at the side of the roadway.

LOCOMOTIVE NAMES

Appropriately, the naming of a number of locomotives reflected the history of the S&DR and of Darlington as a railway town. For example, Edward Pease was honoured in his own day by the S&DR itself. The 2–4–0 engine No. 114 was named *Edward Pease* in 1856, just two years before he died. Other S&DR pioneers who had the honour of engines named after them were [Robert or George] *Stephenson*, No. 65; [Timothy] *Hackworth*, No. 71; [Thomas] *Meynell*, No. 115; and *John Dixon*, No. 174. One suspects that members of the Pease family, who at all times were active on the board of the S&DR, were influential in these name allocations, as also in the choice of *Darlington* for two engines, No. 19 and, when that was sold, No. 103. The story of how a later L&NER locomotive came to be named *Darlington* is intriguing. In 1936 the brand new B17 'Football' class 4–6–0 No. 2852 was about to enter service. North Road workers who built her were disappointed that none of these new express engines was to be named after their local team so, instead of fitting the officially designated *Sheffield Wednesday* name-plates, they made their own *Darlington* plates, which were fitted quietly by night shift workers. So impressed were the L&NER directors by North Road's workmanship in making the plates that they allowed them to stay on No. 2852, and No. 2851 became *Sheffield Wednesday* instead.[10]

Darlington has been perpetuated in the naming of Inter-City 125 power-car No. 43110 in August 1984. It must be conceded that *Stockton* preceded *Darlington* as a locomotive name by some years when S&DR No. 4 (replacing the original No. 4 *Diligence*) was named after the town in 1841.

LASTING IMPRESSIONS

There is one very distinctive feature by which we may recognise a railway town, and that is the necessary noise of the railway system at work. It is not one sound but many, which blend to provide a continuous, accepted background to the community's daily life. As one of the larger railway towns, Darlington contributed those sounds in good measure. First, there was the familiar sound of steam engines at work, a combination of shrill whistles, chimney blasts, exhaust steam and coupled wheels pounding iron and steel tracks. There were the sounds of passenger trains on the main lines – slow branch line trains and expresses, some of which sped past Bank Top Station on the avoiding line yet all the same made their presence felt by the pitch of the noise of their passing. There were the noises of the goods trains on the slow lines as they came to a halt and each wagon buffered up to its neighbour, and the staccato sound echoed all along the formation to the guard's van at the end, followed by the noise of couplings taking the strain as the train set off again when the signal arm fell to announce the 'all clear'. There were sounds, similar and yet with subtle differences, emanating from the wide marshalling yards as wagons were shunted by diminutive tank engines, or from innumerable locomotives going about their business on trip workings. These sounds, of steam escaping, of metal on metal, were augmented from time to time by the shrill sirens that marked the beginning and the end of working shifts. Given the nature and the variety and the amount of all this noise, there was no mistaking that Darlington in its heyday was indeed a railway town.

There were also many visible signs of the industry's presence within the wide railway corridors bisecting the town. (A good vantage point was the footbridge, which at one time spanned the entire width of the tracks to the north of Bank Top Station.) There were the busy lines themselves with their signal gantries and overbridges, coaling stations and water towers. There were buildings and plant by the side of the line which could belong to no other industry – signal-boxes, covered engine sheds and roundhouses open to the sky, turntables and repair shops, point levers and ground frames, cattle pens and loading bays, lettered and numbered signs by the track-side for the information of drivers and firemen. There were larger buildings also – two passenger stations with cavernous train-shed portals, goods depots and coal yards, engineering departments and administrative offices. Even away from the railway corridors there were – and in many cases still are – reminders of the industry's presence. In the early days there were the company's offices in the town centre. Later, the four-square bell tower of St John's Church and the soaring clock-tower of the main-line station, like a campanile, came to dominate the skyline from their eminence at Bank Top, where they are still visible from all around.

There are many other tangible relics and reminders, on a smaller scale, of Darlington's history as a railway town; these are dealt with in the following chapter. By all the qualifying criteria, therefore, we may fairly claim that Darlington was for many long years, and despite the recessions of recent times, as much the epitome of a railway town as she was, historically, the first.

> As meteor-like her fiery engines flew,
> O'erleaped the bourn or pierced the mountains through –
> We claim for Darlington the railway mead.
> She formed the nucleus, she foretold the speed.[11]

CHAPTER 11

Heritage

In 1920 the firm of A.E. Berry, pawnbrokers, of 58–61 Northgate, Darlington, produced ornamental metal plaques for sale as souvenirs, which depicted *Locomotion* and her tender, complete with brazier tail-lamp, on a plinth labelled 'No. 1 Engine', together with Bulmer's Stone similarly mounted and labelled bringing up the rear. On a slip of paper pasted on the back of the example on display at the Darlington Railway Museum there is printed a plea: 'Preserve the relics of the past: they are the history of the nation'. There could be no better definition of heritage.

Much of the rich heritage of the S&DR which remains today has already been detailed in the previous chapters. That includes the infrastructure (track-bed, buildings by the lineside such as stations, goods depots and engineering works), and real estate further afield (such as the Pease family's home and company offices). On the ground it is still possible to seek them out by consulting some of the more enduring products of the 175th anniversary celebrations – the now rather faded signposts, plaques and information boards erected along the routes of four historic rail trails: 1. from Witton Park to Shildon; 2. in Shildon township; 3. in the Borough of Darlington; 4. in the Borough of Stockton-on-Tees. In this final chapter we concentrate on the locomotives and rolling stock that have been preserved, and the legacy in art and design of the S&DR.

LOCOMOTION NO. 1

The very first locomotive to haul a train on a public railway has providentially been preserved. It is now on view at the Darlington Railway Museum. The manner in which *Locomotion No. 1* acquired its name and number calls for comment. When she was placed on the line at Aycliffe on 10 September 1825, and on the opening day, she was described simply as 'the Company's locomotive engine'. Indeed, the S&DR did not name or number any of its locomotives for several years. During this period, each driver, or 'engineer', drove only his own engine and was paid according to the load it hauled. In consequence, the locomotives were referred to by the names of their drivers. James, or Jem Stephenson, the brother of George, habitually drove *Locomotion* so it was customarily called 'James Stephenson's engine'.

The first engine allocated a name was *Royal George*, bestowed by Timothy Hackworth himself when it made its first trip on 29 October 1827, as recorded in his notebook for the following month, December. The next specific mention of a name is under an S&DR sub-committee minute of 21 October 1831, a record that Hawthorn had agreed to make three engines similar to *Majestic*, so that engine possibly had the name given when new in the summer of 1831. By the autumn of that year, therefore, the settled policy was to name at least the new engines. Unfortunately, when precisely the older engines (including *Locomotion*) were first named is not clear. In the National Railway Museum archives, under the date November 1832, is a table that includes another eighteen engines, seventeen of which bear names. The exception is No. 1, to which a name was not allocated. However, in a report by

Thomas Storey, the S&DR's resident engineer, on engine performance over the period 1 January to 1 July 1833, there are twenty-one engines listed by name, and *Locomotion* is among them. This is the first record of the name. The second occurs in a report by the traffic manager, John Graham, dated 13 September 1833.

In none of these records are numbers given against the names. The first record of names *and* corresponding numbers, under the date 9 March 1833, is of four provided under contract by William Lister. Those were No. 3 *Black Diamond*, No. 7 *Rocket*, No. 12 *Majestic* and No. 16 *Director*. Again, *Locomotion* is not among them. Numbers must surely have been given, however, when the third and fourth engines arrived in 1826. The first mention of any number, in a sub-committee report dated 5 October 1827, reads 'Report having been made that no. 2 engine has received grievous injury at Stockton . . .'. The second mention is an entry in one of Timothy Hackworth's notebooks, under the date 25 October 1827: 'Paid 2s. to Thos. Clark for taking the fireman's place who got himself lame at no. 1 locomotive.' This is the first record of *Locomotion*'s number.

There is no contemporary evidence that in these early years numbers were actually painted or placed upon the locomotives themselves. However, under the date 30 May 1831, there is a sub-committee minute: 'Timothy Hackworth is directed to have Nos. with large figures projecting from the chimney of each Locomotive Engine.' It was immediately after this direction, therefore, that the engine we know as *Locomotion* had the inseparable and eponymous number – 1 – fixed to its chimney. At first, numbers were applied to all engines in service (except *Royal George*, which was the only one to carry a name at that time). Later, in February 1839, a decision was taken that coaching (that is passenger) engines should carry names only, and that mineral (that is goods) engines should have both names and numbers.[1] When an engine was moved from the list of mineral engines to the list of coaching engines it ceased to be numbered. According to Samuel Smiles, the purpose of the direction to affix numbers to engines was for identification, and principally so that complaints against drivers for misdemeanours such as excessive speed, causing lineside fires through excessive spark discharge from chimneys, inebriation, or carrying illicit passengers (such as girls) could be more readily and exactly reported – and punished.

During its lifetime the S&DR, unlike other companies, never put its own name or initials on any of its engines or tenders. The combined identification plate for *Locomotion No. 1* was retained for many years, but with subtle variations. For the 1875 Jubilee Exhibition the year '1825' was added to the name. In 1961, when the engine was removed to North Road shops for overhaul and restoration, the number and date were removed and the name alone has remained to this day.

Locomotion's original appearance was not exactly as it is today. For example, the one-piece cast-iron wheels with eight straight spokes with which the engine was first fitted were subject to breakages. They were accordingly replaced with Timothy Hackworth's patented plug wheels. Originally, *Locomotion*'s tender would have contained a barrel to hold water for the engine, and not the riveted iron rectangular water tank resting on the tender sides at the back, which is usually depicted in paintings and incorporated in replicas that purport to represent *Locomotion* on the opening day. In evidence, there is the hand-written reminiscence dated 20 May 1857 of one of the original S&DR employees, Robert Metcalf, recollected in his seventy-seventh year when he lived in retirement at 11 Church Street, Darlington:

> . . . when No 1 engine was put on to yon mount afront the station there was agreat deal discussion about her I could condicked [contradicted] them in many waords but I thought it was not my place to do so she all in an original state excepting the tender it

was a water barrel but on to top on an end on a mucln waggon and she travled as nigh as I can tell for 2 years before she got a proper tender.

Locomotion has also had a succession of different boilers throughout her life. However, two original features remain that are of interest. A brazier is attached to a swivelling bar pivoted in the centre of the rear end of the tender. When the arm was swung in line with the engine and tender the brazier acted as a tail-lamp. When *Locomotion* was hauling a train of wagons it was swung to the side so as not to obstruct coupling-up. The second feature is a brass bell, inscribed 'S&DR', which was rung to warn of the approach of this unfamiliar invader of an otherwise peaceful scene.

From the opening day on 27 September 1825, *Locomotion* was in regular use on the line until she was withdrawn from service in 1841. On 4 June 1846 she was brought out of retirement to haul the inaugural train on the M&RR. Her return to base at Shildon was probably her last outing under her own steam. Later in 1846, she was being used at Howden Station as a pumping engine – 'fitted up with pumps from the engine at Middlesbrough'. From 1850 to 1856, she was used as a stationary pumping engine at Pease's West Colliery near Crook, owned by Henry Pease and Partners. When, early in 1856, the engine was deemed unfit for further use, she was included in a batch of S&DR engines advertised for sale at an auction to be held at Stockton on 16 January 1856 – incredible as this may seem to us in our heritage-conscious age. Fortunately, the intrinsic historical value of the engine was realised in time. *Locomotion* was withdrawn from the sale and restored as near as possible to her original condition, at a cost of £50, by her owners, Henry Pease and Partners, probably at Pease's West Colliery.

On 20 May in the following year, 1857, *Locomotion* was mounted with her tender on a stone plinth in the open air outside North Road Station – the first time a historic locomotive had been preserved and displayed in public. The plinth carried a terse abbreviated inscription incised in the stonework: 'S&DR No 1. 1825'. (This was the occasion of Robert Metcalf's affidavit recorded above.) In 1892, following a petition from Darlington Corporation and others who were concerned at its deterioration, *Locomotion* was transferred to Bank Top Station by the NER, and placed under cover, at the south end, on a raised plinth at the head of the twin bays serving trains to Saltburn and Richmond. In 1898 the engine was removed and overhauled again at North Road Works while the plinth at Bank Top was extended to accommodate an additional S&DR locomotive, No. 25 *Derwent*. *Locomotion* was installed on the extended plinth, and *Derwent* was placed next to her, on 22 April 1899. Both engines stood on old stone-sleepered track.

During *Locomotion*'s years on 'permanent' display at North Road and Bank Top Stations her fame ensured that she was removed on numerous occasions for exhibition elsewhere, including Philadelphia for the US Centennial (1876); Chicago, USA (1883); Liverpool (1886); Newcastle upon Tyne for the Queen Victoria Jubilee Exhibition (1887); Paris (1889); Edinburgh (1890); and Wembley for the British Empire Exhibition, together with No. 4472 *Flying Scotsman* as the L&NER exhibits (1924). In addition, *Locomotion* took part in the 1875 S&DR Jubilee Exhibition, and in February 1881 in another static exhibition of locomotives in Infirmary Yard, Newcastle upon Tyne, as part of the centenary celebrations of George Stephenson's birth on 9 June 1781. Rather confusingly, two engines named *Locomotion* featured in that exhibition, the other being the L&NWR 2–2–0 No. 1867 built at Crewe in 1852. *Derwent* was also exhibited alongside *Locomotion*. Finally, the engine took part in the 1925 S&DR centenary procession, and the exhibition of static locomotives held at Faverdale, Darlington.

In 1941, during the Second World War, both *Locomotion* and *Derwent* were removed from Bank Top Station and stored for safety in an old S&DR engine shed at Stanhope in Weardale. They were returned at the end of the war and removed again in 1961 to North Road Shops for overhaul and restoration. They were finally replaced on their familiar plinth in June of that year.

On 27 September 1975 *Locomotion* was removed from her plinth for the penultimate time and installed at Preston Park as part of Stockton's celebrations for the 150th anniversary of the S&DR – the sesquicentenary. Later in 1975, it was thought appropriate to transfer both *Locomotion* and *Derwent* to the newly created Darlington Railway Museum in North Road Station as its principal attractions. *Derwent* was in place on the opening day, 27 September 1975, but it was not until a week later, on 4 October 1975, that *Locomotion* was finally installed. In a very real sense, she had come home, for she stands upon, or very close to, the site of the line on which she originally ran in 1825.

For the cavalcade of locomotives and rolling stock that comprised the highlight of the 150th anniversary celebrations in 1975, a replica *Locomotion* was built by engineering apprentices under the direction of the veteran railwayman Mike Satow, as a training exercise, in various centres throughout the north-east. It was subsequently moved to the North of England Open Air Museum at Beamish, County Durham, where it is frequently steamed and run on a dedicated section of track.

In Art and Design

The original Robert Stephenson and Company's working drawings for *Locomotion* have not survived, apart from a simplified drawing developed from an early design sketch by George Stephenson, with notes.[2] In the 1880s Theodore West, chief draughtsman at the North Road Works, published a number of sheets of simple sketches of locomotives showing their development from the earliest days. Two of these sheets depict locomotives of the S&DR and the NER, and *Locomotion* heads the first sheet.[3] A much more accurate sketch of the engine minus its tender, together with nine other 'early Locomotive Engines of the S&DR Company 1825–1862' and a table of their particulars and dimensions, was prepared by Clement Edwin Stretton, chief engineer, to illustrate his popular history *The Locomotive Engine and its Development*, first published in London in 1892.[4] An excellent 'Conjectural Drawing of No. 1 Locomotion – as delivered, and drawings of the same engine as preserved', one of a series by Alan Prior, is reproduced in T.R. Pearce's *magnum opus*, *The Locomotives of the S&DR*.[5]

Apart from such measured drawings, *Locomotion* has inspired practitioners in every field of fine, applied and decorative art. The first known representation was on one of the four large flags that streamed above the inaugural train. Beneath the inscription 'S&DR, Opened for Public Use 27th September 1825' and the company's motto, was a picture of the engine hauling wagons against an appropriate background. A number of paintings that also portray this flag have already been mentioned in the text, including, of course, John Dobbin's 1875 painting of the opening day.

Locomotion has featured on at least seven postage stamps from as many countries. A set of four Great Britain stamps was issued in 1975 to commemorate the sesquicentenary. The 7p stamp depicts the engine and its tender with the captions 'Stephenson's Locomotion' and '1825 Stockton and Darlington Railway'. She has also featured on a railway poster with the caption 'L&NER Centenary 1825–1925 Locomotion No 1'. A number of finely detailed scale models of *Locomotion* are on view to the public in museums, including two in the Darlington Railway Museum. Also on display is an unusual representation of *Locomotion* in relief on a large bronze plaque, measuring 3 ft by 1½ ft, designed and made by Alexander McNay of Darlington in 1885. *Locomotion* and her tender are crisply carved on the end of a pew in the north aisle of Stockton parish church in the High Street, and, on the back, Samuel Smiles' version of the company coach *Experiment* – 'The gift of Mr. A Pallister in 1925'. Finally, *Locomotion* is reproduced in coloured enamel on one of six panels on Stockton's ceremonial mace.

With the possible exception of *Rocket*, *Locomotion* has featured more often, in every form of art, than any other steam engine. There could be no more appropriate symbol of railway history in general, and of the history of the locomotive in particular.

DERWENT

The S&DR locomotive No. 25 *Derwent* was built by W. & A. Kitching and Company at its Darlington works in 1845 (works number 12). The engine still carries its maker's name-plate, 'A Kitching Darlington', an indication that it was built by Alfred Kitching alone after his brother William retired from the firm. *Derwent* was one of Timothy Hackworth's numerous 'Tory' class of 0–6–0 mineral engines and as such was one of the first to be fitted with his return-flue boiler.

After an accident-free period of service with the S&DR, *Derwent* was sold in 1868 to Messrs Pease and Partners to replace No. 12 *Trader*, which blew up that same year. *Derwent* was employed at Pease's West Colliery near Crook, and was also involved in the construction of the Waskerley Reservoir. She accompanied *Locomotion* to the George Stephenson centenary celebrations in Newcastle upon Tyne in 1881, and Queen Victoria's Diamond Jubilee celebrations, also in Newcastle, in 1887. As the sole survivor of Timothy Hackworth's 'Tory' class, she was presented to the NER by Pease and Partners in 1898. She took part in the 1825 centenary procession under her own steam, and is said to have reached 12 mph during a trial run the day before.

Among the employees, including a driver of *Derwent*, Joseph Dobson, who was aged ninety in 1924, this locomotive was always referred to as *The Derwent*, despite the name-plate on the engine. A large number '25' was fixed to the steam dome above the maker's name-plate when the engine was built. This number is fortuitous, since it not only represents the locomotive's official number in the company's records, but it can also be taken to allude to the date of the opening of the S&DR. However, it has to be said that *Derwent* was the second engine to carry this number. The first was an 0–6–0 named *Enterprise*, to a design by Timothy Hackworth, also built by W. & A. Kitching and Company in 1835. It was probably disposed of in 1841. After our now preserved *Derwent* was sold in 1868, the number was transferred to yet another locomotive, an unnamed 0–6–0 designed by William Bouch and built at Darlington North Road Works in 1869. This in turn was disposed of in 1907.

As preserved at the Darlington Railway Museum, with its distinctive 'tender at both ends', *Derwent* appears in a dark green livery. In fact, very little is known with any certainty of the colours used, or when a definite S&DR livery came into being. However, there is a general consensus that green was employed for locomotives possibly as early as No. 23 *Wilberforce* of the 'Director' class, which was received from Hawthorne's in 1832 or 1833. The green in which *Derwent* appears may therefore be similar, or even identical, to this first S&DR livery.

Of course, *Derwent* too has been the subject of numerous photographs and illustrations. It features, for example, in an excellent series of measured drawings by Alan Prior: '0–6–0 Engines and Tenders, by W & A Kitching, for the S&DR, as Preserved at North Road Station, Darlington'.[6]

COMPANY CARRIAGES

We have already remarked on the debt owed by the early railway passenger coaches, in both form and appearance, to their road-running stagecoach counterparts. A contemporary writer tells us: 'The most costly and elegant contain three compartments and resemble the body of a coach [in the middle] and two chaises, one at each end, joined together.'[7] Although only three S&DR coaches have been preserved, dated as they are to different periods of the company's history, they serve to indicate a 'typological development' from the flowing lines of the stagecoach towards the plainer, rectangular, functional style with which we are familiar today. By the 1840s, the first, experimental period of railway carriage construction was reaching its close, most standard-gauge systems in Britain having adopted a three-compartment, four-wheel, leaf-sprung design, of which the first of our examples was a pioneer.

1. The oldest S&DR coach, and indeed the oldest railway passenger coach in the world, is preserved at the Darlington Railway Museum. No. 31 was built in 1846 by Messrs Horner and Wilkinson at their works in Commercial Street, Darlington, at a cost of £230. Although the two ends are squared off, the three compartments are distinguished by gilded strip-mouldings; being 'bowed' below, the impression given is of three stagecoaches joined together, albeit mounted on one continuous iron under-frame. The interiors each have two lateral rows of seats facing each other, with markedly superior, luxurious upholstery and decoration in the first-class compartment. It is evident from the arrangements on this example at the Railway Museum that the coaches were provided with facilities for a guard, who rode aloft in a very exposed position on a seat at one end of the roof, to which he ascended by means of an iron ladder fixed to the rear of the coach. 'From this perch he had to climb down

S&DR early passenger coach No. 31 of 1846, preserved at the Darlington Railway Museum.

to apply the brakes whenever the engine-driver whistled instructions.' Luggage was conveyed outside in a railed-off space in the middle of the roof.

2. The second S&DR coach, in point of date, is on display at the National Railway Museum. No. 59 was built at Carlisle in 1848 for a cost of £280, and saw service until 1872. It resembles No. 31 in most of its principal features, including the rich burgundy livery and the crest beneath the first-class compartment window, but with the serial number 59 in the centre. (It follows that in 1848 the S&DR was operating some sixty passenger vehicles.)

3. The final development of the S&DR four-wheeled coach is represented by No. 179, a four-compartment Third, which was built at the company's Darlington, Hopetown Carriage Works. Its severely rectangular shape is achieved by the use of horizontal planking and an outside frame to the body – a far cry from the curving stagecoach lines of its prototypes in the 1830s. No. 179 was built in 1867 after the company had formally amalgamated with the NER, but until 1873 still continuing as a separate concern.[8] It is known as the 'Forcett Coach' because it saw service between Darlington and the terminus of the nominally independent Forcett Railway. At the time of writing, it may be seen in the Soho Works Construction Shed of the Timothy Hackworth Victorian and Railway Museum at Shildon, awaiting the go-ahead for a full-scale restoration.

All three preserved coaches are now part of the National Collection administered by the National Railway Museum at York.

SCULPTURE

Portraits of the principal characters who shared the eventful history of the S&DR have already been included as illustrations in this account. Some of the more prominent have also been honoured by statues in their memory.

The most significant event of the jubilee celebrations, in view of its lasting consequences, was the decision to commission a statue of Joseph Pease, 'M.P., of Southend, Darlington' and eldest son of Edward Pease, who, as we have seen, was himself one of the prime movers of the S&DR, and therefore of railways in general. The statue was erected in the town centre and on 27 September 1875, fifty years to the day after the opening, after the band of the Grenadier Guards had struck up 'God Save the Queen' and 'Rule Britannia', it was unveiled by Harry Vane, the 4th Duke of Cleveland.

The statue is a full-length figure, by Joseph Lawson of London, and depicts Joseph Pease in the Quaker dress of his day, right arm across his chest, hand thrust into his waistcoat in the style of Napoleon. On the plinth are four bronze plaques depicting some of the manifold interests of this versatile and public-spirited philanthropist. The plaque facing east features a Dame School, symbolising education; west, the figure of an anonymous campaigner, together with a group of slaves, under the slogan 'Freedom', symbolising the anti-slavery movement; south, Joseph Pease himself in the lobby of Parliament with Lord Palmerston, Lord John Russell and other politicians, symbolising his work as a Member of Parliament; and north, featuring a number of industries in which he was involved. In the foreground is a stylised, primitive version of an early locomotive with its tender and a train of wagons, and in the background the masts and rigging of ships, smoking factory chimneys and mills.

The statue was originally placed in the middle of the road at the junction of Bondgate and Northgate, surrounded by iron railings and gas lamps. It has been re-sited on a number of occasions since, when it proved to be an obstruction to traffic, and now stands only a few yards from its original site, at the north end of High Row, facing south by east.

Statue of Joseph Pease, sculpted by Joseph Lawson of London and erected and unveiled in High Row, Darlington on the occasion of the jubilee celebrations of the opening of the S&DR in 1875.

It may be thought regrettable that there is in Darlington no memorial to either of the two Fathers of Railways, George Stephenson and Edward Pease. In the case of Edward Pease, the absence of a memorial in Darlington is the more surprising, given his close association with his native town where he lived all his life and where he died and is buried. It is true, however, that throughout his life, in true Quaker tradition, Edward Pease refused all honours. When, in 1857 (the year before his death) the leading citizens of Darlington proposed to erect a statue in recognition of his services, Edward Pease would have none of it. Instead, on 23 October 1857, he received twenty prominent citizens at his home, led by Francis Mewburn, who proceeded to read out a complimentary address of appreciation, no doubt to the embarrassment of all concerned. It may be that since then no posthumous statue to Edward Pease has been proposed in deference to his wishes. After the lapse of a century and a half, however, it might be fitting to mark the Millennium by realising the desire of his contemporaries, and so honour in our day Darlington's most distinguished son.

Three statues of George Stephenson were commissioned at the height of his fame. The first, of marble, was ordered by the directors of the L&MR at the end of December 1844 to honour their own engineer, the 'Father of the improved Railway of Modern Times'. It was placed in St George's Hall, Liverpool, where it still remains. The second statue, of Carrara marble, 15 ft tall, by Edward Hodges Baily (1788–1867), was ordered posthumously and placed in the vestibule of the L&NWR Station in Euston Square, London, in 1852. When that became the London, Midland and Scottish Railway's Euston Station, the statue was relocated to the foot of the staircase (LM&SR) that led in Philip Charles Hardwicke's noble station-hall to the Shareholders' Room above. When, in the 1960s, Euston station was remodelled, the statue was removed to the north end of the Great Hall of the National Railway Museum, where it stands on a high brick plinth. Stephenson is depicted in his working clothes, grasping a rolled-up surveyor's plan. The third statue, in bronze, by the Northumbrian sculptor John Graham Lough (1798–1876), was erected in 1867 out-of-doors at the junction of Westgate Road, Neville Street and Collingwood Street, in Newcastle upon Tyne, close by Central Station and the former premises of the Newcastle Literary and Philosophical Institute, with which George and his son Robert had a close association. It still remains there, surrounded by traffic. Stephenson is rather improbably clad in what appears to be a Roman toga. Appropriately, there are represented at the corners of the pedestal, as supporters, the seated figures of engineering workers – a pitman, a mechanic, an engine-driver, and a plate-layer.

A bronze statue of Robert Stephenson by Baron Carlo Marochetti (1805–67) was completed in 1861. It was erected outside Euston Station in 1871, where it stands today against a modern glass and concrete block.

On a smaller scale, two porcelain busts, one of George and the other of Robert Stephenson, both 15 in high, were manufactured by Josiah Wedgwood and Sons from sculptures by Edward William Wyon (1811–85). They were issued in 1858, ten years after George Stephenson's death,[9] and only one year before the death of Robert. Both are on display at the National Railway Museum, together with a 3 ft high bust of George Hudson, 'The Railway King 1800–1871'. Two marble busts, each 30 in high, one of George Stephenson sculpted by Joseph Pitts in 1846 and the other of Robert Stephenson sculpted in 1856 by Charles H. Mabey, can be seen at the National Portrait Gallery. A bronze relief profile-portrait of Richard Trevithick was executed more recently by L.S. Merrifield and was fixed to the wall of University College, Gower Street, London, in 1933, the centenary of the railway pioneer's death.

There is only one modern work of art in Darlington which, in the words of its creator, 'reflects Darlington's place in railway *history*' – although, in view of its nature, that claim may be considered a little misleading. On 23 June 1997 a replica locomotive 130 ft long, composed of 181,754 bricks and weighing 1,500 tons, was unveiled on a site in Morton Park next to Morrison's Supermarket, which had commissioned it.[10] The location is alongside Darlington's A66 eastern by-pass road, near its junction with Yarm Road some 1,000 yards south of the now abandoned original line of the S&DR. Although titled *Train*, the sculpture is of a single locomotive loosely based on the record-breaking L&NER 2-6-4 streamlined A4 Pacific No. 4468 *Mallard*. It was designed by the sculptor David Mach, built by Shepherd Construction, cost a total of £760,000, and was put together by a team of engineers, architects and thirty-four bricklayers. The choice of *Mallard* to commemorate Darlington's railway history is not particularly apt since this distinguished locomotive had no closer association with the town other than its regular appearance hauling expresses on the east coast main line.[11]

Models and Memorabilia

Although the numerous models and other memorabilia associated with the S&DR are listed and described in Appendix III, 'Catalogue of Museum Exhibits', a small number call for special comment.

As part of the centenary celebrations in 1925, an exhibition of locomotives and rolling stock, railway equipment and relics, documents and drawings, etc., was organised at

The 'First in the World' medal issued by the L&NER to commemorate the centenary of the S&DR in 1925.

Faverdale. The exhibition was opened by the Duke of York accompanied by the Duchess (the future King George VI and Queen Elizabeth). To mark the event, the Duke was presented by Viscount Grey with a silver model of *Locomotion* on an inscribed base, and the Duchess with a companion silver model of the company's coach *Experiment*. Curiously, in her letter of thanks, which has been preserved, the Duchess mistakenly refers to the model of *Experiment* as a dandy cart. (Both models are on display at the Timothy Hackworth Victorian and Railway Museum. Significantly, the replica train included the slab-sided 'prison-van' version of *Experiment*, not the 'stagecoach' version, and this is what is reproduced in the model.)

The Faverdale exhibition, which lasted until 18 July, was visited by the delegates to the International Railway Congress on 3 July, for whom a banquet was provided. The Italian delegates presented a large bronze shield commemorating George Stephenson to the L&NER, which was accepted on behalf of the company by William Whitelaw, its Chairman (1922–38). (Italy's first railway line was constructed under the supervision of George Stephenson's son Robert.) The shield measures 3.1 m by 4.1 m and weighs approximately 1 ton. It features two Italian railway workers bearing aloft a portrait medallion of George Stephenson, and, at the foot, a representation of *Locomotion* and a contemporary L&NER engine. The shield now stands outside the National Railway Museum at York.

Another enduring reminder of the centenary celebrations is the medal, briefly referred to in the Introduction, that was issued by the L&NER to mark the occasion. The obverse is inscribed 'FIRST IN THE WORLD 1825–1925' and features a railwayman holding in his right hand a model of *Locomotion* and her tender on a section of track, with No. 4475 *Flying Scotsman* in the background. The reverse is inscribed 'EDWARD PEASE: GEORGE STEPHENSON. STOCKTON & DARLINGTON RAILWAY INITIATED 1825. LONDON NORTH EASTERN RAILWAY 1854–1925.'

Earlier, in 1871, a medal had been issued to commemorate the centenary of the birth of George Stephenson. The obverse carries a legend round the circumference which reads 'GEORGE STEPHENSON CENTENARY MEDAL', and features a representation of *Locomotion* with her tender. Oddly, although smoke issues from the funnel, the engine is depicted on the plinth she occupied at the time, inscribed 'No. 1. 1825', above the caption: 'THE FIRST LOCOMOTIVE ENGINE THAT EVER HAULED A PASSENGER TRAIN'. The reverse has the legend 'GEORGE STEPHENSON. BORN JUNE 9. 1781, AT WYLAM, NEWCASTLE-ON-TYNE. DIED AUG. 12. 1848.' It features the head and shoulders of a mature George Stephenson in profile.

Banknote

Perhaps the most universal, artistic and tangible reminder of the S&DR in general, and of *Locomotion* in particular, is the design of the current £5 banknote, first issued in 1990. With a portrait of HM Queen Elizabeth II and the figure of Britannia on the obverse, its principal feature on the reverse is 'a large portrait of an historical figure from British history' – George Stephenson. There are sixteen other design items referring to the work of George Stephenson and the railways he pioneered. Not unexpectedly, the S&DR is prominent among them. *Locomotion* is depicted with the inaugural train crossing Skerne Bridge on the opening day, preceded by the outrider holding aloft the flag that warned pedestrians of its approach. Between this horse rider and the engine, the name *Locomotion* is repeated over and again in a radial pattern. The view through the arch of the bridge is of a stationary winding engine-house taken from the corporate seal of the S&DR – a case of artistic licence. The scene is combined with a side elevation of *Rocket* in the foreground above the twin captions: 'STEPHENSON'S ROCKET LOCOMOTIVE 1829' and 'SKERNE BRIDGE ON STOCKTON and DARLINGTON RAILWAY 1825'. A third caption, 'GEORGE STEPHENSON 1781–1848', is written beneath the portrait in a script similar to that of his customary signature.

In the centre of the reverse is a geometric pattern representing the sun, from which a ray of light hits the ground just below *Rocket*, producing flames. This represents the first lighting of the boiler of *Locomotion* when placed on the track at Aycliffe Lane. Among numerous items of railway interest on the obverse is an intaglio circular pattern based on Timothy Hackworth's invention, the plug wheel.

A Future for the Past?

Although some of the visible evidences of the S&DR in its home town have disappeared, an encouraging number still remain. Most of them could appropriately be gathered under the headings 'Street Furniture' and 'The Everyday Environment'.

The Old Town Hall and its companion Market Hall were designed by Alfred Waterhouse, built at a cost of £16,000, and opened in May 1864 in good time for the founding of

Darlington as an independent municipal borough in 1867. The clock-tower, to the north of the market and which balances the Town Hall to the south, was also designed by Waterhouse and came into use in July 1864. It was paid for by Joseph Pease as a gift to his native town. The cast-iron balconies of the Town Hall and the clock-tower are embellished with reproductions of railway engine plug wheels. This motif is exactly copied in cast-iron brackets and benches, respectively, over and on the pavement outside the Dolphin Centre facing Market Square, which was opened in 1983, and on the spandrels of the bus shelters in the town centre.

As Darlington moves into the twenty-first century, those four affirmations that had their beginnings in the nineteenth century and which have characterised the S&DR and its home town in the course of this account, carry even greater force:

> *FLOREAT INDUSTRIA* – LET INDUSTRY FLOURISH
> *OPTIMA PETAMUS* – LET US SEEK THE BEST
> PRIVATE ENTERPRISE FOR PUBLIC SERVICE
> FIRST IN THE WORLD

George Stephenson died in his sixty-seventh year on 12 August 1848 at his home, Tapton House, near Chesterfield. He was accorded the almost unprecedented honour, for a layman, of burial beneath the altar of Holy Trinity parish church (newly built in 1837 in a medieval Gothic style). The spot is marked by a ledger stone simply inscribed 'G S 1848'. In front of the altar there is a much larger stone slab inscribed:

> GEORGE STEPHENSON 1781–1848 RAILWAY PIONEER
> FIRST PRESIDENT OF THE INSTITUTION OF MECHANICAL ENGINEERS

The Victorian stained-glass east window behind the altar was erected to the memory of George Stephenson, and his entwined initials are incorporated several times in the design. Finally, nearby, on the south wall of the sanctuary, there is a plain and unadorned tablet with the following inscription: 'In Memory of ELIZABETH, Wife of George Stephenson of Tapton House, who died Aug 3rd 1845 Aged 66 Years. And Also of the Above named GEORGE STEPHENSON Who Died Aug:12: 1848 Aged 68 Years'.[12]

Edward Pease survived George Stephenson into a vigorous old age. When aged eighty-eight, he was described by Samuel Smiles as 'hale and hearty, full of interest in the present, with a bright eye and the mental vigour of a man in his prime, and with an elasticity in his step which younger men might have envied'.[13] After a short illness he died on 31 July 1858 in his ninety-second year. He was buried on 6 August in the graveyard behind the Friends Meeting House in Skinnergate, Darlington. The place of his burial is marked only by a simple, unpretentious headstone.[14] (It is the custom of the Quakers that all gravestones should be identical, carrying only a name, age, and date of decease, in the belief that all are levelled in death, whatever their eminence in life.) Just before he died, and despite his age and some infirmity, the venerable 'Father of the Railways' was able to travel to Stephenson's home to say farewell to his old friend and partner. He recorded the event in his diary in his characteristic Quaker style:

> *Wednesday, Aug. 16.* Left home in company with John Dixon to attend the interment of George Stephenson at Chesterfield, and arrived there in the evening. When I reflect on my first acquaintance with him and the resulting consequences my mind seems almost

Memorial stone to George Stephenson and his wife Elizabeth in the chancel of Holy Trinity Parish Church, Chesterfield.

lost in doubt as to the beneficial results – that humanity has been benefited in the diminished use of horses and by the lessened cruelty to them, that much ease, safety, speed and lessened expense in travelling is obtained.

Of the many obituaries of George Stephenson that appeared in the press, the *Derby and Chesterfield Recorder*'s tribute of 18 August 1848 rings most true:

What faults he had (and who would pretend that he was without them), cease to be remembered now that he is no more. Take him for all in all, we shall not look upon his like again. Nay, we *cannot*, for in his sphere of invention and discovery, there cannot again be a *beginning* . . .

It is appropriate – and no derogation of the great man himself or of his obituary – to adapt that last sentence and apply it to his temporary home, for in the sphere of steam-hauled public railways there cannot again be a beginning. In the incomparable railway enterprise, Darlington was then, and therefore will always remain – FIRST IN THE WORLD.

APPENDIX I
Pease Family Homes in Darlington

(OTHER OWNERS IN PARENTHESIS)

THE BULL later THE GROVE	JOSEPH & MARY PEASE, *née* Richardson	1763–1821
146 NORTHGATE	EDWARD & RACHEL PEASE, *née* Whitworth	1798–1858
EAST MOUNT	JOHN & SOPHIA PEASE, *née* Jowett	1823–
ELM RIDGE	JOHN & SOPHIA PEASE, *née* Jowett	–68
FEETHAMS	JOSEPH & (1) ELIZABETH PEASE, *née* Beaumont (2) ANNE PEASE, *née* Bradshaw	
NORTH LODGE	JOHN BEAUMONT & SARAH PEASE, *née* Fossick	1833
BORROWSES later SOUTHEND	(EDWARD BACKHOUSE) JOSEPH & EMMA PEASE, *née* Gurney Daughters EMMA & JANE[1]	1800 1826–72 1872–95
PIERPONT later PIERREMONT	(JOHN BOTCHERBY) HENRY & (1) ANNA PEASE, *née* Fell (2) MARY PEASE, *née* Lloyd[2]	1830 1845–1909
WOODBURN	SOPHIA, *née* Pease & THEODORE FRY	1868–
WOODLANDS	(ROBERT BOTCHERBY Jnr) SIR JOSEPH WHITWELL & MARY PEASE, *née* Fox[3] (SIR WILLIAM LEE)	1815– 1854–61 1902
BRINKBURN	HENRY FELL & ELIZABETH PEASE, *née* Pease WALTER FELL PEASE	
MOWDEN HALL	EDWIN LUCAS PEASE & FRANCIS HELEN, *née* EDWARDS	1881–1927
WOODSIDE	(JOHN HARRIS) GURNEY & KATHERINE PEASE, *née* WILSON	1840– –1915
HUMMERSKNOTT	ARTHUR & MARY LECKY PEASE, *née* Pike	1864–
VILLA	HERBERT PIKE PEASE	

Other Quaker Homes in Darlington

POLAM HALL	HARRINGTON LEE	1794–
	JONATHAN & HANNAH BACKHOUSE	1825–50
	EDMUND BACKHOUSE	
	THE MISSES PROCTOR	
WEST LODGE	THOMAS BACKHOUSE	1798–
	DAVID & ANNIE DALE, *née* Backhouse	1853–
PARADISE, later LARCHFIELD	JOHN BACKHOUSE	1811–
	FRANCIS & ELIZABETH MEWBURN	1831–84
BELLE ROSE later HOLLYHURST	GEORGE ALLISON	
	WILLIAM HARDING	
ELMFIELD	WILLIAM BACKHOUSE	1815–
	ALFRED KITCHING	
BELLE VUE	JOHN DIXON	–1865
BROOKSIDE	THOMAS MACNAY	

Notes
1 Family moved to Cliff House, Marske by the Sea
2 Family moved to Stanhope Castle
3 Family moved to Hutton Hall, Guisborough

APPENDIX II
Select Genealogy of the Pease Family

APPENDIX II

Family members referred to in the text are given in bold.

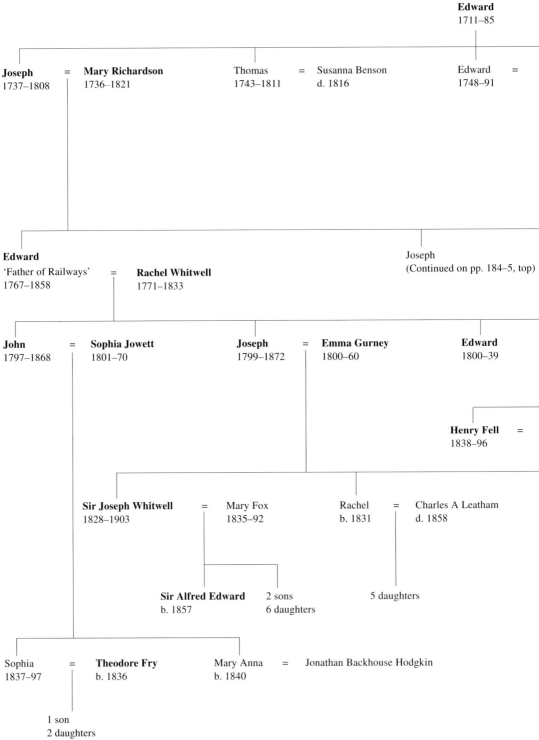

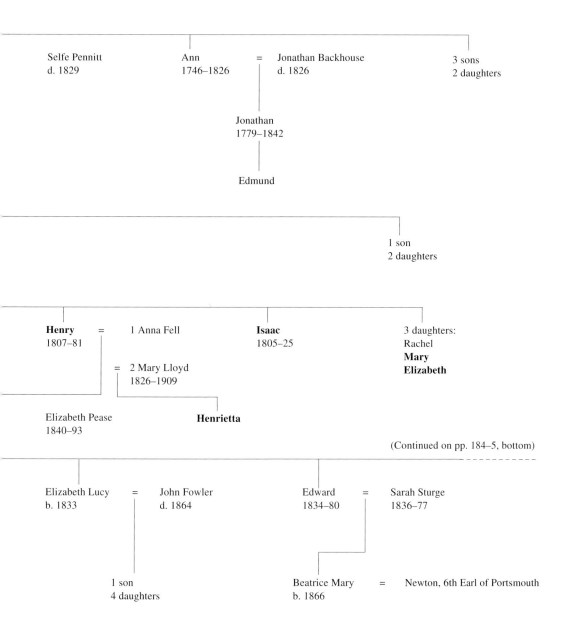

APPENDIX II

Continued

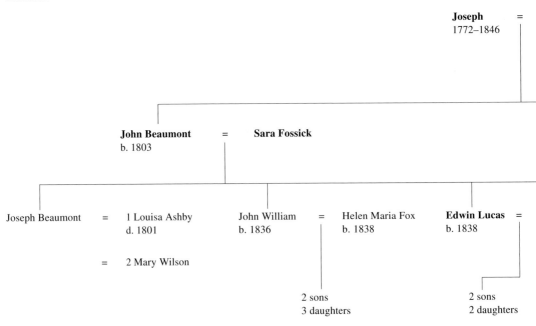

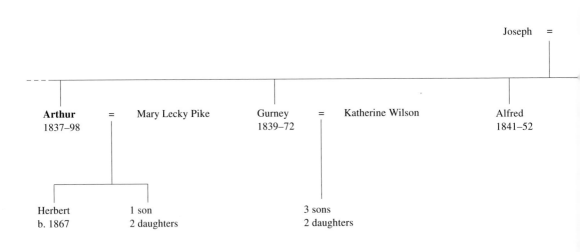

1 **Elizabeth Beaumont**
d. 1824

= 2 Anne Bradshaw
1783–1856

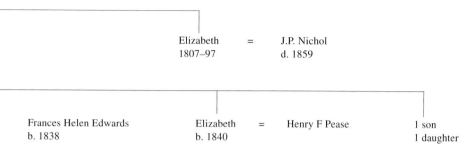

Elizabeth	=	J.P. Nichol
1807–97		d. 1859

Frances Helen Edwards	Elizabeth	=	Henry F Pease	1 son
b. 1838	b. 1840			1 daughter

Emma Gurney

Charles	=	Sara Elizabeth Bewley		2 sons:
b. 1843		b. 1844		John Henry
				Francis Richard

APPENDIX III
Catalogue of Museum Exhibits

Key
NRM   National Railway Museum, York
DRM   Darlington Railway Centre and Museum
THM   Timothy Hackworth Victorian and Railway Museum, Shildon
BOM   Beamish Open Air Museum, Durham

Locomotives

DRM	*Locomotion No. 1*
BOM	*Locomotion No. 1* (replica constructed by Mike Satow and apprentices)
DRM	*Derwent*

Company Carriages

DRM	No. 31 1846, built by Horner and Wilkinson at Darlington
NRM	No. 59 1848, built at Carlisle
THM	No. 179 1867, built at Darlington Hopetown Carriage Works, the 'Forcett Coach'

Goods Wagons

NRM	Dandy cart
NRM	Chauldron wagon from South Hetton Colliery, marked '155 S.H. 2:18'
THM	Chauldron wagon

Models

Locomotion No. 1

NRM	1:8 scale. Length 2 ft 9 in. Working model made for the 1875 anniversary celebrations. Presented to Robert Stephenson and Company. Acquired 1997
NRM	Length 1 ft 6 in
DRM	1:12 scale. Made by apprentices of Taylor Woodrow Plant Co. at Darlington in 1975
DRM	Length 1 ft 6 in
THM	Length 2 ft 5 in. Working model made by B.R. Hunt
THM	Silver 1925 railway centenary royal presentation model on stand

Derwent

DRM*	Length 4 ft 0 in

Experiment – Box Version

THM	Silver 1925 railway centenary royal presentation model on stand

Experiment – Stagecoach Version

NRM	1:6 scale. Length 1 ft 8 in. Labelled 'First Class', 'As Used in 1828'
DRM	Length 1 ft 10 in
THM	Length 1 ft 8 in. Working model made by B.R. Hunt

Panorama

NRM	The S&DR Main Line from Witton Park Colliery to Stockton Quay
DRM	The S&DR Main Line from Brusselton Summit to Stockton Quay

APPENDIX IV
Locomotive Profiles

Locomotion No. 1

0–4–0 First engine of the 'Locomotive' class.
Delivered: 16 September 1825
Cost: £500
Overhauled:
(1) by Pease and Partners, 1856
(2) at North Road works, 1898
(3) at North Road works, 1961

Alternative figures are the result of changes and modifications carried out during *Locomotion*'s working life.

Weight	Tons	Cwts
Engine in running order	8	8
	7	15
Do: Light without fuel or water	6	16
	6	10
Tender in running order	2	17
Do: Light without fuel or water	1	2
(Water 1 ton, coal 15 cwt)		
Total		
In running order, locomotive & tender	11	5

Dimensions

<u>Vertical Cylinders</u>	9 in diameter × 24 in stroke
	10 in diameter × 24 in stroke
<u>Boiler</u>	12 ft length × 4 ft diameter (10 ft 3 in inside ends)
Flue Tube*	11 ft length × 2 ft diameter (58 sq ft surface)
Return Tubes (a)	11 ft length × 1 ft 4 in diameter
(b)	11 ft length × 2 ft diameter
Heating Surface	16,585 sq in
Pressure	50 lb per square in
<u>Wheels</u>	4 ft diameter nominal
	3 ft 11 in diameter actual
Wheelbase	5 ft 5 in
<u>Power</u>	
Tractive Force	1,822 lb
Horse Power	WP psi 25
Adhesive Factor	10

* The single flue was at one time replaced by a flue with two smaller return tubes, then by a double (return) flue, and finally back to the original type of straight-through flue.

Derwent

0–6–0 mineral engine 'Tory' class
Delivered: 24 November 1845
Cost: £1,160. According to Alfred Kitching, the enhanced cost was because of a prolonged period of building from May 1842 to November 1845. It included the ordering and preparation of material at the beginning, corrections, and final adjustments at the end.

Overhauled:
(1) At North Road Works for the 1825 centenary celebrations. Work completed May 1825
(2) At North Road Works, 1961

Dimensions

Vertical Cylinders	15 in diameter × 24 in stroke
	15½ in diameter × 24 in stroke
Boiler	*c.* 14 ft 3 in length × 4 ft 6 in diameter
Multitubular boiler	
99 Tubes each	11 ft 9½ in length × 2 in diameter
Heating Surface	99,576 sq in
Pressure	70 lb per sq in
Wheels	4 ft diameter
Wheelbase	2 × 4 ft 6 in

APPENDIX V
Companies Promoted and/or Absorbed by the S&DR

Name	Period of Independent Existence	First Absorbed by
Bishop Auckland & Weardale	1837–47	Wear Valley
Cockermouth, Keswick & Penrith	1861–1923	London Midland & Scottish
Darlington & Barnard Castle	1854–8	
Eden Valley	1858–62	
Frosterley & Stanhope	1861–3	
Great North of England	1836–46	
Middlesbrough & Guisborough	1852–8	
Middlesbrough & Redcar	1845–58	
Redcar & Saltburn	1858–62	
Shildon Tunnel	1842–7	Wear Valley
South Durham & Lancashire Union (Barnard Castle–Tebay)	1857–62	
South Durham & Lancashire Union (Barnard Castle–West Auckland)	1859–62	
Stanhope & Tyne	1834–42	Pontop & South Shields
West Section: Wear & Derwent	1842–7	Wear Valley
Wear Valley	1845–58	
Weardale Extension	1844–7	Wear Valley

(After Tomlinson, pp. 778–9)

APPENDIX VI
S&DR: Acts of Parliament

Royal Assent	Designation	Authorising
19 April 1821	First Act, 2 Geo IV c44	Stockton to Witton Park Colliery and Yarm branch
		Darlington Depots branch
		Black Boy branch
		Evenwood Lane branch
23 May 1823	Second Act, 4 Geo IV c33	To alter main line and deviate Black Boy and Darlington branches, and to construct Croft branch
17 May 1824	Third Act, 5 Geo III c48	To abandon Evenwood Lane branch and substitute Haggerleases branch
23 May 1828	Fourth Act, 9 Geo IV c60	Bowesfield Junction to Middlesbrough
13 July 1849	Fifth Act, 12/13 Vic c54 (Consolidation Act)	Reincorporation of the S&DR
		Diversion to a line alongside the Leeds Northern line at Egglescliffe
		Acquisition of Middlesbrough Dock
20 May 1851		Additional Capital
28 May 1852		Additional Capital
3 July 1854	Act, 17/18 Vic c115	Tunnel branch
		(Shildon Tunnel North to West Auckland)
16 July 1855		Stooperdale Curve
		Waskerley (Burnhill) Deviation
		Wear Valley Junction Curve (not proceeded with)
		Stanley branch
23 July 1858	Act, 21/22 Vic c116	Amalgamation of: Stockton & Darlington, Wear Valley, Middlesbrough & Redcar, Middlesbrough & Guisborough, and Darlington & Barnard Castle Companies
		Fieldon Bridge to Bishop Auckland
		Improvements at Crook
		Redcar to Saltburn
		Rifts House branch (not proceeded with)
13 August 1859		Hartburn Curve
15 May 1860		Additional Capital
30 June 1862	Act, 25/26 Vic c54	Tow Law to Crook Sunniside Deviation
30 June 1862	Act, 25/26 Vic c106	Amalgamation of: Stockton & Darlington, South Durham & Lancashire Union, and Frosterley & Stanhope Companies
13 July 1863	Act, 26/27 Vic c122	'An Act amalgamating the NER Company and the S&DR Company'

Notes

The following abbreviations are used in the notes.
PRO   Public Record Office
DRO   Durham Record Office
DCRO   Durham County Record Office
NYRO   North Yorkshire Record Office
Kirby   References cited are from Maurice W. Kirby, *The Origins of Railway Enterprise* (Cambridge, 1993)

Chapter 1

1. For a first-hand account of his life and work, see Sir Alfred Pease (ed.), *The Diaries of Edward Pease* (London, 1907).
2. For early canal schemes in the north-east, and County Durham in particular, see William Tomlinson, *The North Eastern Railway: Its Rise and Development* (Newcastle upon Tyne and London, 1915), pp. 33–9.
3. *Ibid.*, p. 39; Kirby, p. 25.
4. Tomlinson, *The North Eastern Railway*, p. 40.
5. Kirby, p. 27.
6. As evidenced in PRO RAIL 667/870, a letter dated 6 December 1818 from Joseph Pease to Jonathan Backhouse. Kirby, p. 194, n. 39.
7. L.T.C. Rolt, *George and Robert Stephenson: The Railway Revolution* (London, 1960), p. 61.
8. After the wars in Europe, the local economy had suffered recession, in common with the country as a whole. In the case of County Durham, this was exacerbated when the local Hollingsworth banking organisation failed. In 1818 a member of the S&DR Committee wrote: 'We lost 12 million sterling by the fall of the County bank in this neighbourhood in 1813.' In these circumstances, capital for new ventures was not forthcoming and it is not surprising that Rennie's Report was shelved.
9. Although the various canal schemes came to nothing, the information they provided, in the form of engineer's reports and plans, calculations of costs and revenues, possible alternative routes and the identification of the various collieries, industries, commercial firms and persons likely to support such projects was of considerable value to the promoters of the S&DR.
10. PRO RAIL 667/150. Kirby, p. 193, n. 20.
11. DCRO Q/D/P 5/2(b). All the early S&DR plans were confirmed as deposited, together with the date, in a handwritten note on the plan itself, by 'John Dunne, Deputy Clerk of the Peace for the County of Durham'.
12. A plan *was* deposited with the Clerk of the Peace of Durham County on 3 April 1819 (DRO Q/D/P 4/1) under the title 'Plan of the Intended Railway from Stockton to the Collieries in the neighbourhood of West Auckland'. Unlike other depositions, it does not carry the name of the engineer or surveyor who carried it out, and it may well be the outcome of the survey undertaken by Robert Stevenson. Significantly, the line crosses Darlington by way of Northgate Bridge and Cockerton. The line is very convoluted at its western end where it serves the collieries in the neighbourhood of West Auckland.
13. DCRO Stockton and Darlington Railway Records, U415j, 37764, *Prospectus: Darlington and Stockton Railway: Observations in Parliament* (Sess. 1821). Kirby, p. 195, n. 56. See also U415j R. 23986, Thomas Meynell (Chairman), *A Further Report of the intended Rail or Tram Road from Stockton by Darlington to the collieries with a branch to Yarm* (February 1821).
14. PRO RAIL 667/856, Richard Miles to George Overton, 20 October 1818. Kirby, p. 194, n. 27.
15. DCRO Pease Family Records U418e PEA, 46/19, Joseph Pease Jr to W. Aldam and T.B. Pease, 26 December 1818. Kirby, p. 194, n. 34.
16. N.C.L. Tomlinson Collection, vol. 1, Lord Darlington to George Overton, 12 July 1819.
17. Kirby, p. 34.
18. Francis Mewburn, *The Larchfield Diary: Extracts from the Diary of the Late Mr Mewburn, First Railway Solicitor* (London and Darlington, 1876), p. 8.

19 PRO RAIL 667/153, 'Report of Sub-Committee of the Darlington and Stockton Railway' (1819). Kirby, p. 194, n. 43.
20 Mewburn, *The Larchfield Diary*, p. 71.
21 DRO Q/D/P 5(a) and 5(b).
22 DRO Q/D/P 7.
23 PRO RAIL 667/897, Leonard Raisbeck to Richard Miles, 30 March 1821. Kirby, p. 195, n. 54.
24 Tomlinson, *The North Eastern Railway*, p. 69.
25 For the purposes of consolidation in the Fifth S&DR Act of 13 July 1849, the previous Acts were repealed, the original company was dissolved, and a new one created. The S&DR and all the branches of the dissolved company were vested in the new, and the several provisions of the original Act relating to the name and the seal were carefully restated: 'The former Proprietors shall be instituted into a Company by name "The S&DR Company" and by that name shall be a body corporate and have a common seal.'
26 DCRO Stockton and Darlington Railway Records, U415j, Henry Pease, 'George and Dragon' (1823). Kirby, p. 194, n. 52.
27 Kirby, p. 37.
28 Tomlinson, *The North Eastern Railway*, p. 69.
29 Alan Russell, *Yarm and the First Railway* (Yarm, 1990), p. 5.
30 Kirby, p. 6.
31 *Ibid.*, pp. 7–8.
32 Reproduced in Peter W.B. Semmens, *Stockton and Darlington: 150 Years of British Railways* (London, 1975), pp. 24–5. Later share certificates were less elaborate in their wording – see Tomlinson, *The North Eastern Railway*, p. 115; John H. Proud, *The Chronicle of the Stockton & Darlington Railway to 1863* (Hartlepool, 1998).
33 The first use of the term 'director' occurs in the Fifth S&DR Act of 13 July 1849: 'The present Directors of the Managing Committee of the S&DR shall be the first Directors of the new Company, viz.: Thomas Meynell the elder, Thomas Meynell the younger, Joseph Pease, John Pease, Henry Pease, Thomas Richardson, Francis Gibson, Henry Pascoe Smith, John Castell Hopkins, Henry Stobart, William Kitching, Alfred Kitching, and Isaac Wilson.'
34 High Pit and West Moor Collieries were close to each other. Prior to his appointment at High Pit, and while he was working at West Moor, George Stephenson bought a one-room cottage at The Three Houses, West Moor, so that on transferring to High Pit he had no occasion to move house. However, the 'single room and garrett' was rebuilt and extended until it became the roomy cottage we see today.
35 *The Story of the Life of George Stephenson* (London, 1859), Smiles, Samuel pp. 150–1. Nicholas Wood participated in all George Stephenson's early locomotive experiments and was the author of the influential *Treatise on Railroads* (London, 1825). Rolt, *George and Robert Stephenson*, pp. 52–3.
36 Pease, *Edward Pease*, p. 97. The mural painting is the fifth of a number round the walls of the Children's Library room which illustrate the history of Darlington. It carries the caption: 'George Stephenson (barefoot) and Nicholas Wood at the Bulmer Stone before going to interview Edward Pease 1821. Presented by Dr. J.T. Sinclair, OBE, MCRS.' The arms of Stockton and Darlington are depicted at the top together with their respective mottoes: *Fortiter et Spes* and *Floreat Industria*. At the foot, *Locomotion* with her tender hauling a chauldron wagon is depicted crossing Skerne Bridge.
37 Pease, pp. 85–6.
38 This document is reproduced and published for the first time as No. 9 in *Archive Teaching Unit No. 11: The Stockton and Darlington Railway 1825*, edited by S.C. Dean and R.M. Gard (University of Newcastle upon Tyne School of Education, 1975).
39 Pease, *Edward Pease*, p. 86, quoted in Smiles, *Life of*, p. 151.
40 Rolt, *George and Robert Stephenson*.
41 DCRO Hodgkin Papers, D/140/C/63, 'Thomas Richardson'. Edward Pease to Thomas Richardson, 10 October 1821. Kirby, p. 195, n. 60.
42 Quoted in Kirby, pp. 39–40.
43 Other examples of George Stephenson as prophet have been recorded. Even in his Killingworth days he forecast, in conversation with old Robert Summersides: 'I will do something in coming time which will astonish all Englishmen.' He was true to his word. Equally unlikely at the time was his forecast that he would live to see the mails carried by steam power from London to his native Newcastle. 'I will send the locomotive as the great missionary over the World,' he declared – and so he did.
44 It was Dr Dionysius Lardner, the scientific populariser, who first described George Stephenson as 'Father of the locomotive' in 1836.
45 William Hylton Dyer Longstaffe, *The History and Antiquities of the Parish of Darlington* (Darlington, 1854 and 1973), p. 363.

46 There is an excellent portrait of Francis Mewburn – sculpted in relief, head and shoulders in profile – in the Darlington Railway Museum. It carries a caption that reads: 'Electrotype Bronze. 30th November 1785 – 11th June 1867. Presented by the Mayor and Councillors. Originally in the Old Town Hall.'
47 DRO G/D/P 8. The plan is reproduced in Tomlinson, *The North Eastern Railway*, p. 66.
48 Randall Davies, *The Railway Centenary, A Retrospect* (L&NER, London, 1925), pp. 18–19. Although he prefaces this extract with the words 'Mewburn notes in his diary', it is not found in *The Larchfield Diary*.
49 The figures are quoted from the original documents transcribed in the *Northern Echo, Railway Centenary Supplement*. A copy of the Report in the S&DR Records gives the following figures: 'Old £79,781 11*s* 2*d* New £64,637 13*s* 3*d*.' In Joseph Priestley's *Historical Account of the Navigable Rivers, Canals and Railways Throughout Great Britain* (London, 1831) he quotes £84,000 for Overton and £74,300 for Stephenson. See Rolt, pp. 71–2, fn.

Chapter 2

1 PRO RAIL 667/8, Stockton and Darlington management committee minutes, 23 July 1821. Kirby, p. 195, n. 61.
2 Rolt, *George and Robert Stephenson*, p. 71.
3 Quoted in Davies, *Centenary*, p. 13.
4 Rolt, *George and Robert Stephenson*, p. 69. John Dixon had a brother, James, resident in Darlington. *Ibid.*, p. 174.
5 C.A. McDougall, *The Stockton & Darlington Railway, 1821–1863* (Darlington, 1975), p. 10.
6 Rolt, *George and Robert Stephenson*, p. 72.
7 See section 'Lineside', Chapter 4, p. 00.
8 Rolt, *George and Robert Stephenson*, p. 71.
9 M. Heavisides (ed.), *The History of the First Public Railway, Stockton and Darlington: The Opening and What Followed* (Stockton-on-Tees, 1912), pp. 35–6.
10 In a marginal comment on his personal copy of *Observations on the Proposed Railway or Tramroad*, 1818.
11 Tomlinson, *The North Eastern Railway*, p. 71.
12 Kirby, p. 36.
13 Michael Longridge and John Birkinshaw were the joint authors of *Remarks on the Comparative Merits of Cast Metal and Malleable Iron Railways, and an Account of the Stockton and Darlington Railway* (Newcastle, 1827). (See also n. 14, below.)
14 DCRO S&DR Records U415j; Michael Longridge and John Birkinshaw, 'Remarks on the Comparative Merits of Cast Metal, and Malleable Iron Rails', Bedlington, 28 February 1821. Kirby, p. 195, n. 65.
15 PRO Rail 667/8, S&DR sub-committee minutes, 1821–1836, 29 December 1821. Kirby, p. 195, n. 66.
16 Smiles gives what is probably an embroidered and imaginative account of Stephenson's recommendations and evidence to the directors, and of the questions they put to him. George Stephenson recommended wrought-iron rails with more confidence 'since they had been in use at Killingworth for eleven years. The waggons pass over them daily, and there they are, in use yet, whereas the cast-iron rails are constantly giving way.' Smiles, *Life of*, pp. 159–60.
17 The doubling of the main line entailed the provision of 987 tons of malleable-iron rails, 285 tons of cast-iron chairs, and 70,000 stone blocks for sleepers. McDougall, *The Stockton & Darlington Railway*, p. 27.
18 P.J. Holmes, *The Stockton and Darlington Railway, 1825–1975* (Ayr, 1975) adds: 'Where, at the edge of a plateau, loaded waggons worked downwards – instead of a *winding engine* a *wheel* was provided, so that the weight of loaded waggons attached to one end of the rope pulled the empty waggons at the other end.'
19 F = ma - Force = mass times acceleration.
20 On the S&DR network as a whole '60 great ropes, of a total length of 68 miles, were daily travelling over these stationary engines and self-acting inclines at a speed of from 7 to 11 miles an hour'. Tomlinson, *The North Eastern Railway*, p. 16.
21 'The switches at "meetings" worked automatically, one set of waggons throwing them into position for the passage of another.' *Ibid.*, pp. 376–7.
22 Rolt, p. 93.
23 Rastrick's rough notes were copied at Newcastle on 24 January 1829. Each engine had cylinders of 21 in diameter and 3 ft 3 in stroke. The 'Small Barrel' (Drum) was 5 ft diameter and the 'Large Barrel' 10 ft 4 in diameter. Rastrick and Walker reported in favour of fixed engines, much to the chagrin of George Stephenson. The pair of 15 hp engines were later upgraded to one of 40 hp (commissioned on 26 January 1831), and later still to a pair of marine engines.
24 There is an archival photograph of the complex as it appeared about 1900 in the Durham Record, No. 09139, Darlington Library Accession No. PH 7321, 21-91A.

25 Holmes, *The Stockton and Darlington Railway*, p. 25.
26 Reproduced as Pl. III in Tomlinson, *The North Eastern Railway*, p. 16.
27 Carl von Oeynhausen and H. von Decken, 'Report on Railways in England in 1826–27', trans. E.A. Forward, *Transactions of the Newcomen Society*, vol. 29 (1953–5), pp. 1–12.
28 The wagons were brought to a standstill at this point. 'A great deal of the excess of power acquired by the descending waggons was employed in taking the ascending waggons over the "kip", and this velocity being reduced by the additional effort they were the more effectively brought to a halt.' Tomlinson, *The North Eastern Railway*, p. 377.
29 The 'cow' was first used at Whitehaven in 1765, and at Killingworth many years before the opening of the S&DR. *Ibid.*, pp. 380–1, and notes.
30 Robert Young, *Timothy Hackworth and the Locomotive* (London, 1923), p. 111. There is a similar print on the same page, taken from a watercolour, of the Etherley engine-house as it appeared in 1875. It clearly shows a double line of stone sleeper blocks still *in situ*.
31 Tomlinson, *The North Eastern Railway*, p. 380, quoting John Glass, *A Railway Engineer of 1825* (1875).
32 The full title is Robert Stephenson and Joseph Locke, *Observations on the Comparative Merits of Locomotives and Fixed Engines as Applied to Railways, Compiled from the Reports of Mr George Stephenson* (Liverpool, 1830).
33 Tomlinson, *The North Eastern Railway*, p. 376.
34 The gradients on these inclines were as follows: Etherley south and north 1:33; Brusselton west 1:30, east 1:33.5; Mount Pleasant, Stanley (1858) south 1:16–1:9, north 1:26; Crawley 1:12–1:8; Weatherhill 1:21–1:13; Park Head 1:86–1:82; Meeting Slacks 1:47–1:35; Nanny Mayor's 1:14; Sunniside (1858) 1:19–1:13; Howden (1843) 1:44.
35 The question is sometimes asked, why was the company registered in the name of Robert Stephenson, and not that of his father, George? The answer has to do with the complex but extremely close relationship between father and son – two very different personalities. A quite separate company, George Stephenson and Son, was formed to execute railway surveys and construction. Rolt, *George and Robert Stephenson*, pp. 100, 103.
36 Michael Longridge seems to have shared George Stephenson's gift of prophecy (see Chapter 1, n. 40 herein). In a letter to Robert Stephenson dated 17 August 1824, Longridge wrote: 'I have little hesitation in saying that the mode of conveyance on Rail Roads by steam will soon become National. You will readily conceive what a vast source of business to us (the Forth Street Works) they will prove if we do but rightly manage it.'
37 Thomas Richardson and George Stephenson each contributed £500. Smiles, *Life of*, p. 158. See also Kirby, p. 45 and n. 77.
38 'Originally so-called because it stands by the side of what used to be called the old pack-road between Newcastle and Hexham.' Smiles, p. 15.
39 According to Samuel Smiles, the engine was run backwards and forwards on a temporary way laid down at the Pipewellgate Foundry and never left the works, but was dismantled from the wheels and set to blow the cupola of the factory. Smiles, *Life of*, p. 89. Its prototype, the Pen-y-Daren engine, was similarly used to drive a tilt hammer at Homfrey's Ironworks.
40 Stanley Mercer, 'Trevithick and the Merthyr Tramroad', *Transactions of the Newcomen Society*, vol. 26 (1947–9), pp. 89–103. The event is captured in an atmospheric painting by Terence Cuneo in the Collection of the National Museum of Wales.
41 A drawing of the scene in Euston Square by Thomas Rowlandson, and a photograph of a model of 'Catch-me-who-can', are reproduced in Anon., *The New £5 Note and George Stephenson* (Loughton, 1990), fig. 4, p. 7.
42 Rolt, *George and Robert Stephenson*, p. 125.
43 Hedley had previously applied to Trevithick for employment at Pen-y-Daren but, perhaps fortuitously, had been declined.
44 The Wylam Colliery 0–4–0 *Puffing Billy* is the oldest locomotive in the National Railway Collection and is on display in the Science Museum, London. An early photograph is reproduced in Davies, *Centenary*, p. 24.
45 'Popularly called Blu*tc*her'. That is either a mistake, or a simple transliteration of the local rendering of 'Blucher'. Smiles, *Life of*, p. 100. Whereas Smiles has Stephenson saying to Pease 'Come over to Killingworth and see what my *engines* can do' (*Ibid.*, p. 152), Sir Alfred Pease substitutes the singular *Blutcher*. Pease, *Edward Pease*, p. 86. It is probable that the locomotive was not directly named after the popular Prussian General Blucher, an ally of Wellington who came to his aid at the Battle of Waterloo, but indirectly from one or both of two other sources: 1. a public house, also named for the General, which was a popular place of resort for the local enginemen; 2. when George Stephenson was working at High Pit his father, old Robert Stephenson, was working at *Blucher* Pit, part of the Killingworth complex close by.
46 'In this respect, if in no other, George Stephenson had set in motion the first locomotive of modern type.' Rolt, *George and Robert Stephenson*, p. 50.

47 Smiles, *Life of*, pp. 131–3; figure, p. 132; Rolt, *George and Robert Stephenson*, p. 52.
48 Samuel Smiles mistakenly implies that it was the people of the neighbourhood of the Hetton Colliery Railway who first styled a locomotive 'the iron horse'. Smiles, *Life of*, p. 141.
49 George Stephenson's supposed watercolour of a 'Killingworth Engine' is reproduced in colour in Davies, *Centenary*, p. 28, and in monochrome in Anon., *The New £5 Note*, fig. 3, p. 5. Other drawings are reproduced in Semmens, *Stockton & Darlington: 150 Years*; Rolt, *George and Robert Stephenson*, p. 39; Tomlinson, *The North Eastern Railway*, p. 21.
50 Kirby, p. 47 and n. 88. DCRO Pease Family Papers, U415j, vol. III, memorandum of a meeting held in Robert Stephenson and Co's Office, 31 December 1824. 'By 1844 Longridge had withdrawn from the business.' Kirby, p. 196, n. 88.
51 Young, *Timothy Hackworth*, p. 103. 'There is no proof that Timothy Hackworth was at the Forth Street works at the time.' Rolt, *George and Robert Stephenson*, p. 82.
52 L.T.C. Rolt suggests that this was a spare wad of oakum packing for the feed pump which the fitters at Forth Street thoughtfully provided. 'There is surely some symbolic significance in this little piece of humble and quite spontaneous ritual by which the sun's heat kindled fire in the belly of the first locomotive in the world to move on a public line of railway.' *Ibid.*, p. 83.
53 *Ibid.*, p. 76.
54 £40,000 of this amount was lent by the Quaker Gurney banking partnership, and £20,000 by the bill-brokers Richardson, Overend and Company. PRO RAIL 667/955, 'Interrogation from the Exchequer Loan Board with answers from the Railway company', July 1826.

Chapter 3

1 Reprinted in the *Courier* of 4 October 1823.
2 James Anderson's account in the *Newcastle Courant* of 1 October 1825.
3 Terence Cuneo's painting is now owned by the National Railway Museum. A more recent painting of a similar scene was executed by Alan Tinley to mark the 'Rail 150' Anniversary in 1975 (see Chapter 11). It now hangs in the vestibule of the Council Chamber in Darlington's new Town Hall. The painting depicts a moderately flat rural landscape, with two goats, two startled horses, and cattle, in the foreground, and a stagecoach in the distance.
4 The National Railway Museum displays a very realistic scale model of the entire route from Witton Park to Stockton, and the Darlington Railway Museum a similar model from Brusselton Summit to Stockton. In both cases, the inaugural train of *thirty* vehicles together with *Locomotion* and her tender, preceded by the outrider waving his flag, is frozen in time crossing Skerne Bridge.
5 1825 was the year of a double tragedy for Edward Pease and his wife Rachel, since their eldest daughter Mary died on 30 May aged only twenty-three.
6 Rolt, *George and Robert Stephenson*, p. 195.
7 Mewburn, *The Larchfield Diary*, p. 12.
8 A retrospective exhibition of Dobbin's work was held in 1996 at the Darlington Arts Centre. The excellent illustrated catalogue prepared by Alan Suddes, then Curator of the Borough of Darlington Museum and Art Collections, includes a reproduction of *The Opening of the Stockton and Darlington Railway*.
9 Reproduced in Anon., *The New £5 Note*, p. 13; Jack Simmons (ed.), *Rail 150: The Stockton and Darlington Railway and What Followed* (London, 1975), p. 22.
10 Tomlinson, *The North Eastern Railway*, pl. V, opp. p. 60.
11 First reproduced in an issue of the *Graphic*, 13 October 1888. More recently in Rolt, *George and Robert Stephenson*, pl. 4; Semmens, *Stockton & Darlington: 150 Years*, p. 34; jacket illustration, Dean and Gard, *Archive Teaching Unit No. 11*.
12 Also reproduced in *Archive Teaching Unit No. 21*; Longridge and Birkinshaw, *Remarks on the Comparative Merits*, 1832 edition; and the principal view in Davies, *Centenary*, p. 31.
13 Reproduced in *Archive Teaching Unit No. 31*; Tomlinson, *The North Eastern Railway*, p. 95.
14 New collieries were sunk at Oakenshaw and Brandon by Messrs Straker and Love, at Waterhouses and New Brancepeth by the Pease family, and at Brancepeth by the Elswick Coal Company. Further to the west, new pits were opened out at Coundon, St Helen's Auckland, Hunswick, West Stanley and Newton Cap. Kirby, p. 151 and n. 18.
15 By the autumn of 1827 three more staithes were in operation at Stockton, and 'merchandise' goods were being carried in returning empty coal wagons.
16 Smiles, *Life of*, pp. 164–5.

17 Henry Booth, *An Account of the Liverpool and Manchester Railway* (Liverpool, 1830), p. 69.
18 Smiles, *Life of*, p. 172.
19 Rolt, *George and Robert Stephenson*, p. 36.

Chapter 4

1 Rolt, *George and Robert Stephenson*, p. 157.
2 Kirby, p. 90.
3 *Ibid.*
4 *Ibid.*, p. 110 and n. 34.
5 *Ibid.*, p. 25.
6 *Archive Teaching Unit No. 30.*
7 James Stephen Jeans, *Jubilee Memorial of the Railway System: A History of the Stockton and Darlington Railway and a Record of its Results* (London and Newcastle upon Tyne, 1875 and 1975), p. 79.
8 Smiles, *Life of*, p. 166, and fig. V, later reproduced in the *Illustrated London News*. Tomlinson is highly critical of Smiles' version: 'No reliance whatsoever can be placed on the well-known woodcut in Smiles' *Life of George Stephenson*.' Tomlinson, *The North Eastern Railway*, p. 109.
9 Smiles, *Life of*, p. 169.
10 Reproduced in Pearce, *Locomotives of the S&DR*, p. 122.
11 Engraving, *Ibid.*, p. 123. From Joseph Gladding Pangbourne, *The World's Railway: Historical, Descriptive and Illustrative* (New York, 1894).
12 According to Rolt, the frame arrived from Newcastle on the day of the trial (Rolt, *George and Robert Stephenson*, p. 83). If so, then attaching the body *in situ* must have been accomplished in a remarkably short space of time.
13 Tomlinson, *The North Eastern Railway*, p. 109.
14 Printed in the *Darlington and Richmond Herald*, 13 November 1875.
15 The First Act of 1821 limited the period during which rates of tonnage and tolls could apply. The Second Act of 1823 extended this period for a further two years, 'if the Quarter Sessions justices think fit'.
16 *Archive Teaching Unit No. 23.*
17 McDougall, *The Stockton & Darlington Railway*, p. 28.
18 Tomlinson, *The North Eastern Railway*, p. 127.
19 Smiles, *Life of*, p. 169.
20 Tomlinson states that *Express* was owned by Richard Pickersgill, since one coach only was let to him, and although the sub-committee on 11 November 1825 instructed Richard Otley to approach the various coach-makers in Darlington for a sketch or pattern of a railway coach, there is no evidence that an order was given for it. He concludes that since there is no record of an S&DR order for *Express*, and since one coach only was *let* to Richard Pickersgill, '*Express* was in all probability his own'. Tomlinson, *The North Eastern Railway*, p. 109.
21 The starting point at the Fleece Inn was later superseded by the Bay Horse at the foot of Castlegate in Stockton. *Ibid.*, p. 130.
22 Longstaffe, *History and Antiquities*, p. 365.
23 Rolt, *George and Robert Stephenson*, p. 219.
24 Tomlinson adds: '. . . as well as the Railway Tavern, which the Company were then building, near the Stockton depots'. Could this be the world's first railway-sponsored public house? He also notes that the Black Lion Hotel starting point was later substituted by the Custom House Tavern at the foot of Finchale Street. Tomlinson, *The North Eastern Railway*, p. 127. The *Union* advertisement is reproduced in *Ibid.*, p. 125.
25 From Henry Pease's speech at the S&DR Jubilee banquet, 1875, *Newcastle Daily Chronicle*, 15 January 1875.
26 The subjects of timetables and other aids for passengers are dealt with in the section 'Timetables and Precedents' in Chapter 5, p. 00.
27 PRO RAIL 667/32, S&DR management sub-committee minutes, 20 April 1833 to 4 October 1833. Kirby, p. 201, n. 108.
28 Tomlinson, *The North Eastern Railway*, p. 120.
29 PRO RAIL 667/945, Thomas Storey to management committee, 21 October 1825. Kirby, p. 198, n. 13.
30 *Ibid.*, Thomas Storey to Joseph Pease, 2 November 1825.
31 Tomlinson, *The North Eastern Railway*, p. 159.
32 The rival claims are detailed in a letter from George Stephenson to Timothy Hackworth dated 25 July 1828.

33 'Dandy' was a common north-east expression for anything new or 'fancy'. A dandy coach ran on the Brampton Railway (illustrated in Tomlinson, *The North Eastern Railway*, p. 528). A dandy cart complete with a life-size model horse is on display in the main hall of the National Railway Museum. 'The origins of this exhibit are uncertain, but it was either constructed as a replica, or restored at the Shildon works of the NER.' Photograph, Semmens, *Stockton & Darlington: 150 Years*, p. 31. Figure in Tomlinson, *The North Eastern Railway*, p. 157. There is an excellent print of a dandy cart to a scale of 1½ in to 1 ft in the Darlington Railway Museum. It incorporates three elevations and a plan together with a sketch of the cart attached to a train of eight chauldron wagons. The caption is incorrect in claiming that the dandy cart was in use on the S&DR from the opening on 25 September 1825 until 1841. Reproduced in McDougall, *The Stockton & Darlington Railway*, p. 40, and *Archive Teaching Unit No. 25*.

34 Longstaffe, *History and Antiquities*, p. 365, citing George Head, *A Home Tour through the Manufacturing Districts of England in the Summer of 1835* (London, 1836).

35 It was a common practice for fare-dodgers to make use of dandy carts. Tomlinson, *The North Eastern Railway*, p. 364.

36 PRO RAIL 667/31, minutes of the Stockton and Darlington Railway sub-committee, 11 November 1825. Kirby, p. 198, n. 20.

37 *Ibid.*, 18 November 1825. Kirby, p. 198, n. 21.

38 *Ibid.*, 25 November 1825. Kirby, p. 198, n. 22.

39 Not to be confused with William Creed. In Robert Metcalf's account of his early life he mentions some of the early drivers and firemen: 'James Stephenson and William Creed firemen and James Stephenson engine driver Robert Mires did not come for a month or two after line was open out when Manchester and Liverpool line was open out William Creed whent to run Mr Hackworth's engine and he never came down here more.'

40 Stephenson and Locke, *Observations on the Comparative Merits*, p. 11.

41 Storey was ordered to keep the most careful records of coal per ton-mile for haulage by the locomotives and the fixed haulage engines. His figures for 1826 were (cost per ton-mile): Locomotives $\frac{2}{25}$ of a ¼d; Fixed Engines – Brusselton 1⅛d, Etherley 1⁴⁄₁₆d. The figure of 30 per cent savings when comparing locomotive with horse haulage coincides closely with that of ¼d per ton-mile which he had earlier estimated on the basis that: 'One locomotive Engine will perform 6 Stockton and Yarm journeys and 3 Darlington journeys each weekday at 45 Tons each journey or 405 Tons per week, and at 50 weeks per annum is equal to 20,250 Tons. The cost of one engine, to work it, upkeep it, coal consumption and watering and interest on capital is £373 2s per annum and will lead during that time 45 tons 7,100 miles which is equal to ¼d per ton per mile.' Rolt, *George and Robert Stephenson*, pp. 129–30.

42 The only known drawing of *Experiment* occurs in J.U. Rastrick's notebook. It is a rough sketch that does not show boiler mountings and other details. Reproduced in Rolt, *George and Robert Stephenson*, p. 131. There is a detailed description in Oeynhausen and Dechen, 'Report on Railways in England in 1826–27', p. 3.

43 Rolt maintains that it was not the performance of *Royal George* alone which persuaded the company not to abandon steam traction in favour of horses. 'It has already been shown that the Company had proved the superiority of locomotives over horses before the *Royal George* [which began work in November 1827] appeared.' Rolt, *George and Robert Stephenson*, p. 143.

44 S&DR management committee to Timothy Hackworth, 18 July 1828, 5 September 1828. Young, *Timothy Hackworth*, p. 169.

45 Kirby, p. 68. PRO RAIL 667/3, Stockton and Darlington management committee minutes, 14 July 1829. Kirby, p. 199, n. 53.

46 The other two contenders for the priority of inside cylinders and a coupled axle were *Liverpool* and *Planet* of the Liverpool and Manchester Railway. Rolt has pointed out that while it has always been assumed that the name of the engine *The Globe* referred to its spherical copper steam-dome, there was a public house in Stockton also called The Globe, much frequented by engine drivers. Rolt, *George and Robert Stephenson*, pp. 137–8. *The Globe* was an individual design and did not inaugurate the production of a class of locomotives. Her working life ceased in 1838 when the boiler exploded.

47 Rolt, *George and Robert Stephenson*, p. 175.

48 It has been suggested that the precipitating factor was the discovery of 'fraudulent behaviour' on the part of some of the haulage contractors, in particular the under-recording of journeys in order to evade toll charges. Kirby, p. 94, and n. 107.

49 Tomlinson, *The North Eastern Railway*, p. 176.

50 *Archive Teaching Unit No. 32*.

51 These early signals are illustrated in McDougall, *The Stockton & Darlington Railway*, p. 52.

52 Tomlinson, *The North Eastern Railway*, pp. 411–12, figure, p. 412.

Chapter 5

1. Tomlinson, *The North Eastern Railway*, p. 364.
2. The First S&DR Act of 1821 was quite clear that 'Passage upon the Railway [was] to be free on payment of tonnage'.
3. The facade carries a plaque that reads: 'Pease House. Edward Pease Known as "The Father of the Railways" Lived Here'. It is presently occupied by a firm of Chartered Accountants.
4. Pease, *Edward Pease*, p. 67.
5. Smiles, *Life of*, p. 157.
6. The painting became part of the Flatow Collection. (Louis Victor Flatow (1820–67), art dealer, commissioned Frith's *The Railway Station*.) The two daughters of Edward Pease standing behind the embroidery frame as they are taught by George Stephenson would be Mary and Elizabeth, aged twenty-one and twenty respectively, when this incident took place. The plan of the S&DR in the painting partly obscures another, which is headed '. . . and Newcastle', yet no railway with 'Newcastle' incorporated in its title had been promoted by 1823.
7. The protagonists of No. 156 as the home of Edward Pease cited this painting to support their case, on the grounds that the parlour background to the painting was very similar to a ground-floor back room of the Domestic Science College at the time.
8. On a map of 1826 (reproduced in George J. Flynn, *Darlington in Old Photographs: A Second Selection* (Stroud, 1992), p. 22), No. 9 is shown as occupied by the bank of William Skinner, Esq.
9. Pictured in Tomlinson, *The North Eastern Railway*, p. 189.
10. Reproduced in *Ibid.*, p. 417.
11. Parliamentary Papers, 1839, X, 332.
12. In the S&DR (Consolidation) Act of 1849 limits were imposed on charges for 'small packages and single articles of great weight for any distance', according to their weight, viz.: 14 lb, 6*d*; 14–28 lb, 2*s*; 28–56 lb, 2*s* 6*d*; above 56 lb 'The Company may demand any sum they think fit'.
13. Such as Ralph Smith of Darlington, who altered and improved the original *Experiment*. According to Tomlinson he also altered the post-coach *Eclipse* to run on the railway. Tomlinson, *The North Eastern Railway*, p. 130.
14. Revd James Adamson, *Sketches of our information as to rail-roads. Also, an account of the Stockton and Darlington Railway, with observations on railways, etc.*, p. 57. First printed in the *Caledonian Mercury* and subsequently published in Newcastle in 1826.
15. Adamson, *Sketches of our information as to rail-roads*.
16. The Revd Adamson seems to have been unaware that this was substantially the same brake-mechanism that was employed on the chauldron wagons which preceded railway passenger coaches.
17. Passenger Duty was removed from all fares up to and including 1*d* per mile by the Cheaper Trains Act of 1883, which effectively exempted most third-class fares. In the event, after the passing of the Act, all third-class fares came down to 1*d* per mile. Most lines retained 'Parliamentary', all-stations services until the early years of the British Rail era.

Chapter 6

1. PRO RAIL 667/1158, testimonial for Timothy Hackworth by William Patter, Walbottle Colliery, 3 February 1825.
2. Jeans, *Jubilee Memorial*, p. 268.
3. Kirby, p. 64.
4. *Ibid.*, p. 66, Table 2. Enginemen's earnings and comparative haulage performance, 1828 (citing Tomlinson, *The North Eastern Railway*, pp. 150, 151).
5. Kirby, p. 68, Table 3. Tonnages conveyed and operating costs, May 1828 (citing Timothy Hackworth's memorandum book, in Young, *Timothy Hackworth*, p. 168).
6. Rolt, *George and Robert Stephenson*, p. 132.
7. *Newcastle Daily Chronicle*, 28 September 1875.
8. Tomlinson, *The North Eastern Railway*, p. 397.
9. Young, *Timothy Hackworth*, pp. 312–21.
10. Kirby, p. 25 and n. 21.
11. 'As far as it is possible to judge.' Gilkes and Bouch were also appointed as sub-contractors to the company, with Bouch remaining in that capacity until the amalgamation of the S&DR with the NER in 1863. Kirby, *Men of Business*, p. 108.
12. Timothy Hackworth's headstone has been retained in the graveyard of Shildon parish church. In 1998 an area in the centre of the town, giving access to the church, was tastefully laid out as a pedestrian precinct in his

NOTES

memory. At its centre is a 6 ft lifelike statue of Hackworth by the Barnsley sculptor Graham Ibbeson, protected from the weather by a 'bandstand' roof. The statue was unveiled on 7 September 1998. The plinth carries a simple inscription: 'Timothy Hackworth Pioneer Railway Engineer Shildon 1786–1850'. Shildon as a whole is justly proud of its favourite son. There are many visible evidences of his popularity, including a Timothy Hackworth Primary School, Timothy Hackworth public house, Timothy Hackworth Street and Close, and not least the Timothy Hackworth Victorian and Railway Museum.

13 The last steam locomotive, a British Railways Standard 2–6–2T, was turned out in 1957. The last locomotive as such was a Bo Bo Sultzer 1250 hp diesel engine No. D 7597. Altogether, 2,775 steam and diesel locomotives were built at the North Road Shops between 1863 and 1966.

14 The site occupied by Morrison's covers that part of the North Road Shops which were demolished in 1978/9 after a disastrous fire. Another area became a timber yard and boat builders, and yet another an indoor bowling centre, opened in March 1980.

Chapter 7

1 'By May 1826 the Company was in the grip of a liquidity crisis.' Kirby, p. 57, citing T.E. Rounthwaite, 'An Outline History of the Stockton & Darlington Railway', *Railway Observer*, vol. 26 (1956), p. 37.
2 From October 1952 a large and unsightly locomotive scrapyard was sited alongside this branch until the line was closed in 1964.
3 On both Overton's and Stephenson's plans this branch is shown as terminating across the Yarm–Stockton road because, at this point, the present A135 has the effect of straightening the road, which formerly, as the turnpike, curved to run east through Egglescliffe village, now avoided by the A135.
4 The precise purpose of the coal drop ramps is uncertain. It has been suggested that they were provided so that coal from chauldron wagons on the Black Boy branch could be conveniently unloaded by gravity straight into larger wagons from the main line below for onward transhipment.
5 A colliery was developed a century later near the site of the proposed terminus – Coundon 'Jawblades' Colliery, 1920s–30s.
6 A number of pits were sunk in the general area of the Black Boy branch after its closure, including 'Jawblades' and Auckland Park.
7 At various dates during the nineteenth century these engines were advertised for sale (one of the pair, for example, in 1869), to be replaced by more up-to-date equipment. In March 1867 one engine was transferred to Page Bank near Spennymoor. In 1874 both engines were ordered to be dismantled, but it is not certain if this was carried out.
8 Tomlinson, *The North Eastern Railway*, p. 377.
9 Those who are of the opinion that the original Black Boy branch did not extend beyond the right-angle bend concede that there was a rail link to Coundon Grange Pit at a later date.
10 This branch followed the line of the road to Eldon Lane. Nearby earthworks to the east belonged to a later parallel line opened in 1919 from a junction with the South Durham Railway.
11 PRO RAIL 667/466, 'Considerations Reflecting the Situation of Black Boy Colliery in the event of the Branch Railway not being laid to same', not dated, but about June 1826; PRO RAIL 667/3, Stockton and Darlington management committee minutes, 26 January 1827 and 28 June 1827. Kirby, p. 199, n. 56.
12 It has been suggested, rather cynically, that when the parlous financial straits of the S&DR were sufficiently improved for the proprietors to contemplate building one of the three branches authorised by Acts of Parliament, the Black Boy line was chosen in preference to the Croft and Haggerleases branches because the joint owner of the Black Boy Colliery (with Joshua Ianson) was Jonathan Backhouse, 'who happened to be Treasurer of the S&DR at the time'. Tomlinson, *The North Eastern Railway*, p. 139.
13 Kirby, p. 139. See PRO RAIL 557/1122/1150. Kirby, p. 205, n. 18.
14 Tomlinson, *The North Eastern Railway*, p. 139.
15 Longstaffe, *History and Antiquities*, p. 367.
16 DRO Q/D/P 9/1. Reproduced, with amendments, in the form of an engraving by William Miller of Edinburgh, in *Archive Teaching Unit No. 15*.
17 Rolt, *George and Robert Stephenson*, p. 78.
18 *Ibid*.
19 On the 1861 OS map these collieries were augmented by five more.
20 For an account of the complicated construction of skew bridges as later practised by Robert Stephenson on the London and Birmingham Railway, see Rolt, *George and Robert Stephenson*, p. 248.
21 Figure in Tomlinson, *The North Eastern Railway*, p. 185, and see fn.

22 Pease, *Edward Pease*, entry for 18 August 1828. Kirby, p. 77.
23 Maurice Kirby notes that the S&DR was not formally involved in this development. 'Extra capital borrowing was unacceptable in the light of the existing debt structure and further share issues were to be reserved for the construction of a network of branch lines.' Kirby, p. 79.
24 Longstaffe, *History and Antiquities*, p. 364.
25 Tomlinson, *The North Eastern Railway*, pp. 187–8; figure, p. 188.
26 A plan, elevation and perspective view of Brown's design, drawn by John Dixon, is reproduced in *Ibid.*, pl. VIII, facing p. 186. John Dixon's brother James was resident engineer for the building of the bridge.
27 Longstaffe, *History and Antiquities*, p. 364, quoting Head.
28 On 27 December 1830 the inaugural train was stopped so that the official party could inspect the suspension bridge. Although it was to prove a grave disappointment, it had the distinction of being one of only three suspension bridges in Britain to carry a main-line railway. The second was I.K. Brunel's Wye Bridge at Chepstow (1852) and the third his Royal Albert Bridge at Saltash (1859).
29 At the time of the amalgamation of the S&DR with the NER in 1863 more than one-third of the 160 engines handed over by the S&DR had been built by Gilkes, Wilson and Company.
30 Kirby, p. 134 and n. 7.
31 Tomlinson, *The North Eastern Railway*, p. 529.
32 DCRO S&DR Records, D/XD/35/6, report to shareholders, 7 August 1839. Kirby, p. 203, n. 47.
33 PRO RAIL 667/11, Stockton and Darlington management committee minutes, 14 December 1838 and 15 December 1838. Kirby, p. 203, n. 50.
34 Tomlinson, *The North Eastern Railway*, p. 437
35 Kirby, p. 117 and n. 52 – PRO RAIL 667/222, Middlesbrough Dock leasing agreement, 1841. The lease was confirmed in a S&DR Act of 19 May 1851 ('Conversion of loan to share capital').
36 The Act was titled 'The S&DR (Consolidation of Acts, Increase of Capital, and Purchase of the Middlesbrough Dock) Act, 1849. It noted that the S&DR had advanced a payment of £60,000 towards the construction of the dock, and that the owners were willing, in consideration of that payment, that the dock should become vested in the company. The line of railway from the junction with the Middlesbrough Extension line to the dock was also vested in the S&DR at the same time.
37 G.A. North, *Teesside's Economic Heritage* (Cleveland, 1975), p. 12, and see Tomlinson, *The North Eastern Railway*, p. 51. A plan titled 'The Town of Middlesbrough, the New Dock, and Railway' is reproduced in *Ibid.*, p. 436.
38 J.W. Ord, *The History and Antiquities of Cleveland* (Middlesbrough, 1846), p. 107.
39 Asa Briggs, *Victorian Cities* (Harmondsworth, 1968), p. 241.
40 According to Longstaffe, *Locomotion* (to which he applies the now-discredited name *Action*) was accompanied by the celebrated 'A' engine. Longstaffe, *History and Antiquities*, p. 367. See also n. 11 in Chapter 11 below.
41 Kirby, p. 120; Holmes, *The Stockton and Darlington Railway*, p. 38.
42 Pease, *Edward Pease*, concluding remarks, December 1857. Kirby, p. 148.
43 Pease, *Edward Pease*, p. 89.

Chapter 8

1 Mewburn, *The Larchfield Diary*, p. 125.
2 John Glass, 'The Opening of the Shildon Tunnel', *North-Eastern Railway Magazine*, vol. 3, no. 29 (May 1913), pp. 103–4.
3 A small stone building just to the east of the signal-box at Shildon Junction on the north side of the tracks is said to date from the S&DR period.
4 This accounts for a provision in the Fifth S&DR (Consolidation) Act of 13 July 1849 which removed the extra toll of 6*d* per ton on coal passing over inclined planes if, in the case of the Black Boy branch, 'it also passes through Shildon Tunnel'.
5 The line is so labelled on the 1861 OS map. In January 1828 Christopher Tennant published a map of his proposed 'direct line' to Stockton – the so-called 'Deanery Branch' – which shows an extension of this South Durham Railway line from Eldon, linking with the Clarence Railway at a junction just to the east of Simpasture. If it had been built it would have by-passed the stretch where the Clarence Railway was subject to punitive restrictions as it passed over S&DR metals.
6 The Derwent Iron Company was established in 1837. Some limited deposits of iron ore were discovered in 1839 by John Nicholson at nearby Shotley Bridge. The company collapsed in 1857 as a result of the failure of the Northumberland and Durham District Bank. It was largely kept afloat in 1858 by the action of the S&DR

and the NER, since both had a vested interest in its survival. The techically efficient and highly profitable firm that emerged after it was floated on the Stock Exchange in 1864, after a protracted period of reconstruction, was the Consett Iron Company. The reconstruction was in large part financed by members of the Pease family, in particular Joseph Whitwell Pease. Kirby, p. 163 and nn. 51, 52, 53.

7. At one time this remote and isolated junction was a veritable railway town. No less than twenty-two engines were stabled here.
8. The arrangement involved a major financial investment by the S&DR in return for the surrender of the way leaves to Messrs Pease, Meynell and Hopkins. Kirby, p. 119.
9. On 13 December 1833 Stephenson wrote to Harrison: 'I sincerely congratulate you on the appointment [as engineer of the Stanhope and Tyne Railway], and more particularly as you will be instrumental in extending the beautiful locomotive line from Shields, and now that I am about to leave Newcastle I am glad in leaving a locomotive engine advocate behind. It is not a little remarkable that in a neighbourhood where this class of engines had their birth enemies to them can be found on every side.' PRO RAIL 1148/1, p. 15.
10. Rolt, *George and Robert Stephenson*, p. 263.
11. Negotiations between the S&DR and the Derwent Iron Company were at first frustrated by the absence of Parliamentary powers. Again, however, under an agreement of November 1844, Messrs Pease, Meynell and Hopkins, together with Henry Pease and Henry Stobart, acquired the Wear and Derwent Railway lease and way leaves.
12. Rowley Station was dismantled and rebuilt at the Beamish Open Air Museum in County Durham.
13. The eastern section of the Stanhope and Tyne Railway retained the name Pontop and South Shields Railway.
14. Care must be taken to distinguish between railways and companies with very similar names, viz. The Weardale Extension Railway; The Wear Valley Railway; The Wear Valley Extension Railway; The Wear and Derwent Railway; and the Wear and Derwent Junction Railway.
15. Tomlinson, *The North Eastern Railway*, pp. 224–5.
16. The building of this section of the line was facilitated by the S&DR's acquisition of a private railway 'Connecting the WER (now part of the WVR) at the foot of Sunniside incline', and the S&DR's rival, the West Durham Railway, at Roddymoor Colliery. The purchase was authorised by the S&DR Act of 16 July 1855.
17. Tomlinson, *The North Eastern Railway*, p. 568, figure, p. 569; McDougall, *The Stockton & Darlington Railway*, p. 31.
18. The tracks between West Auckland and Shildon Junction were still shown in place on the 1861 OS map. A photograph of the Etherley north incline shortly after the track was lifted is reproduced in Tomlinson, *The North Eastern Railway*, p. 108. The chimney of the stationary engine-house can be seen in the distance through the arch of the Etherley Road overbridge. Photograph B 6284 in the Durham Archive.
19. T. Richardson, *History of the Darlington and Barnard Castle Railway* (London, 1877), p. 27.
20. Not the Duke of Cleveland who obstructed the D&BCR at Gainford, but his son.
21. 'A new pattern of mixed traffic was being established, and the chief casualty was the NER.' Kirby, *Men of Business*, p. 162 and n. 43.
22. McDougall, *The Stockton & Darlington Railway*, pp. 33–4.
23. Mary W. Pease, *Henry Pease: A Short Story of his Life* (Darlington, 1897), p. 72.
24. On the 1861 OS map the stretch from Spring Gardens Junction to Barnard Castle is rather confusingly labelled NORTH EASTERN RAILWAY DARLINGTON SECTION (BISHOP AUCKLAND, HAGGERLEASES AND BARNARD CASTLE BRANCH). Haggerleases here refers to a Haggerleases Station on this branch where it crossed Haggerleases Lane, and *not* the original Haggerleases branch and its terminus.
25. David Joy, *A Regional History of the Railways of Great Britain*, vol. 14, *The Lake Counties* (Newton Abbot, 1983), p. 32.
26. It was not this line but an earlier one, the 'Projected Kendal and Windermere Railway' (authorised on 30 June 1854), that provoked William Wordsworth (1770–1850) to write his celebrated diatribe in verse against what he saw as the violation of the Lake District scenery which, paradoxically, as the foremost of the Romantic poets, he had done so much to popularise: 'Is there no nook of English ground secure from rash assault? . . .'
27. The Redhills curve passed *under* the CK&PR before joining it half a mile further west at Redhills Junction.
28. Tomlinson, *The North Eastern Railway*, p. 558.
29. Pease, *Edward Pease*, p. 300.
30. Evidence of J.W. Pease before the House of Commons Committee on the Cleveland Union Railway Bill, 15 May 1858.
31. Tomlinson, *The North Eastern Railway*, figure, p. 560.
32. Joseph Whitwell Pease went on to become MP for South Durham, 1865–85, and for the Barnard Castle Division, 1885–1903. He was created a baronet and became Sir Joseph on 18 May 1882.

Chapter 9

1. 'In 1827 some observations were printed in Ripon on the advantages of the proposal.' Longstaffe, *History and Antiquities*, p. 366.
2. Tomlinson, *The North Eastern Railway*, p. 278.
3. Kirby, p. 124.
4. DCRO Stockton and Darlington Records, U415j, 33061, minutes of the Committee of the S&DR, 30 October 1835. Kirby, p. 204, n. 72.
5. Cited in David Brooke, 'The Promotion of Four Yorkshire Railways and the Share Capital Market', *Transport History*, vol. 5, no. 3 (1972), p. 257.
6. Tomlinson, *The North Eastern Railway*, p. 280.
7. David Brook, 'The Promotion of Four Yorkshire Railways and the share capital market', *Transport History*, vol. 5, no. 3 (1972), p. 261.
8. The Newcastle and Durham Junction Railway bought the Darlington Junction Railway on 23 May 1844 and the Pontop and South Shields Railway on 3 August 1844. The Brandling Junction Railway was bought by George Hudson in his own name on 13 August 1844. It was transferred to the Newcastle and Durham Junction Railway three days later on 16 August 1844, which took possession on 1 September 1844.
9. Earlier, on 24 May, a special train drawn by the locomotive *Cleveland* conveyed the directors from York to Gateshead. McDougall, *The Stockton & Darlington Railway*, p. 30.
10. Rolt, *George and Robert Stephenson*, p. 271.
11. Reproduced in Flynn, *Darlington in Old Photographs*, p. 49.
12. Although the Leeds Northern Railway built the two extra tracks, the S&DR was granted Parliamentary approval for their construction as the 'end user'. This is apparent from the wording of the S&DR Consolidation Act of 1849: 'The Company shall be empowered to abandon a portion of their Railway in the Parishes of Egglescliffe and Stockton on Tees and to make a new line of Railway in lieu of the portion of the Railway so proposed to be abandoned.' The Act proceeded to define the location of the beginning and end of the new line. It also stated: 'Nothing in this Act shall prejudice the rights and powers of the Leeds and Thirsk [*sic*] Railway.'
13. In its time, Yarm has enjoyed the luxury of four railway stations, only one of which (the latest) is located in Yarm itself: 1. the original S&DR station at Yarm Junction; 2. the S&DR Coal Depot station; 3. the LNR station; 4. a new commuter station built by Railtrack 1.4 miles south of the last on the outskirts of the town, where the B1264 road to Worsall crosses the Egglescliffe–Northallerton line. At the time of writing the station is used exclusively by the franchise holders Northern Spirit.
14. Longstaffe, *History and Antiquities*, p. 365.
15. Kirby, p. 85.
16. Tomlinson, *The North Eastern Railway*, p. 237.
17. This was an early example of the way in which rival railways inflicted economic disaster on their competitors during the Victorian and Edwardian eras. Peter W.B. Semmens, *Exploring the Stockton and Darlington Railway* (Newcastle upon Tyne, 1975), p. 18.
18. In order to safeguard their supply of iron ore, the Bell Brothers developed additional mines in the decade 1870 to 1880 at Brotton (Carlin How, Cliff, Huntcliffe and Lumpsey), Skelton Park and North Skelton.
19. The extension from Redcar to Saltburn was not opened until much later, on 1 November 1878.
20. Tomlinson, *The North Eastern Railway*, p. 560.
21. DCRO S&DR Records U415j; letter dated 1 December 1858.
22. Kirby, p. 165.
23. Tomlinson, *The North Eastern Railway*, p. 566.
24. *Ibid.*, p. 125.
25. *Ibid.*, pp. 570–6.
26. Part of the stone abutments of this bridge can still be seen.
27. Brian Bailey, *George Hudson: The Rise and Fall of the Railway King* (Stroud, 1995), p. 64.
28. Kirby, p. 162; Tomlinson, *The North Eastern Railway*, p. 526.
29. It appears that the S&DR's decision to amalgamate with the NER was precipitated by an attempt launched in 1860 by the L&NWR to link the Newcastle and Carlisle Railway with the S&DR network via the projected Newcastle and Derwent Railway. The S&DR was faced with the choice of being absorbed by the L&NWR or the NER and chose the latter. Kirby, p. 167.
30. *Ibid.*, p. 172.
31. *The Darlington and Stockton Times*, 12 August 1863.

32 Joseph Pease noted in his diary for 26 March 1860 that the decision in favour of an amalgamation was a unanimous one on the S&DR board, 'save my brother Henry who thinks we merge our usefulness and lose control'. Mewburn, *The Larchfield Diary*, p. 186.
33 By this time, Francis Mewburn was opposed to the amalgamation and refused to attend the meeting. 'After a conversation with John Dixon he reluctantly agreed.' Pease, *Edward Pease*, p. 183.
34 The figures are aggregated from the mileages given for individual lines in the text. The 2.1 miles of running rights from Parkgate Junction to Croft Junction on the east coast main line of the GNER are not included since they were already part of the NER total at the time of the amalgamation.
35 Henry Pease seems to have had a change of heart on being appointed Chairman (see n. 31 above).

Chapter 10

1 See section 'The Coat of Arms' p. 00.
2 B. Barber, 'The Concept of a Railway Town and the Growth of Darlington 1801–1911: A Note', *Transport History*, vol. 3 (1970), p. 283.
3 *Newcastle Daily Chronicle*, 28 September 1875.
4 Barber, 'The Concept of a Railway Town', p. 284.
5 *Ibid.*, p. 286.
6 *Ibid.*
7 Bailey, *George Hudson*, p. 96.
8 1871 Census, vol. II, cited in Barber, 'The Concept of a Railway Town', p. 287. In 1801 the population was only 4,670. Subsequently it grew to 55,631 by 1911, and to 100,600 by 1 May 1997, when the Urban and Rural Districts combined.
9 Alan Suddes, Introduction to *The Godfrey Edition: Old Ordnance Survey Maps: Darlington 1898* (Gateshead, 1987).
10 As recorded in Anon., *Railway Memories No. 2: Darlington and South-West Durham* (Todmorden, 1990), p. 41.
11 Quoted in Longstaffe, *History and Antiquities*, p. 397.

Chapter 11

1 Pearce is of the opinion that although John Graham obviously knew the numbers, 'whether in fact they were so fitted is another matter'. Pearce, *Locomotives of the S&DR*, p. 6.
2 *Ibid.*, figure 15, p. 30; figure 16, p. 31. Transcript of George Stephenson's notes on pp. 29–30.
3 Theodore West, *The Evolution of the Locomotive Engine* (Darlington, c. 1900). The sheets referred to in the text – 'Types of railway locomotive engines on the Stockton and Darlington and the North Eastern lines, 1825–1848' – are reproduced in Theodore West, 'Types of Railway Locomotive Engines on the Stockton and Darlington and the North Eastern Lines, 1825–1848', *Jubilee of the Railway News* (1914), p. 5.
4 Clement E. Streeton has no less than 106 publications on railway subjects to his name.
5 Pearce, *Locomotives of the S&DR*, figure 14, p. 26.
6 *Ibid.*, figure 65, p. 79.
7 Cited in Rolt, *George and Robert Stephenson*, p. 194.
8 Although this coach was built under NER auspices, it carries the S&DR insignia because the NER allowed an autonomous S&DR carriage sub-committee to continue in being for some years after the amalgamation.
9 The bust of George Stephenson is illustrated on the front cover of Anon., *The New £5 Note*. Another example is on display at the Darlington Railway Museum, together with smaller busts of both George and Robert Stephenson in Parian Ware by Minton.
10 The involvement of Morrison's in commissioning this brick sculpture is appropriate since their other supermarket in Darlington occupies part of the site of the North Road Shops, which closed in 1966.
11 *Mallard*'s 126 mph record-breaking achievement took place on 3 July 1938 during a run on the east coast main line, some distance from Darlington, between Grantham and Essendine. However, Darlington did feature in the epic 'batttle of the gauges', which took place as a result of Brunel's dramatic challenge to put the respective merits of the two systems to practical test in a series of locomotive trials. As Rolt has recounted, the courses selected for the contest were York to Darlington for the standard gauge, and Paddington to Swindon for the broad gauge. In the event, the standard-gauge locomotives *Stephenson* and *Engine A* were severely trounced by their broad-gauge rival *Ixion* of earlier design. Rolt, *George and Robert Stephenson*, p. 293.

NOTES

12 *Ibid.*, p. 297; Smiles, *Life of*, pp. 156, 426, tablet figured on p. 429. Elizabeth Hindmarsh, George Stephenson's second wife, whom he married in 1830, was the daughter of a farmer at Black Callerton. At one time, Stephenson had worked as engineer at the colliery there. His third wife was a Miss Gregory, whom he married six months before he died. She also was the daughter of a farmer, near Bakewell in Derbyshire, and had been his housekeeper.

13 Smiles, *Life of*, p. 149.

14 Close by is the headstone commemorating Edward Pease's wife Rachel, and another that commemorates both Joseph Pease, Edward's son, and his wife Emma, who predeceased him. This part of the graveyard contains orderly rows of headstones above the graves of the many members of the Pease and Backhouse families. (See Appendix II, p. 181.)

Bibliography

Anon., *Railway Memories No. 2: Darlington and South-West Durham* (Todmorden, 1990)
Anon., *The New £5 Note and George Stephenson* (Loughton, 1990)
Adamson, James, *Sketches of our information as to rail-roads. Also, an account of the Stockton and Darlington Railway, with observations on railways, etc.* (Newcastle, 1826, extracted from the *Caledonian Mercury*)
Ahrons, E.L., *The British Steam Locomotive, 1825–1925* (London, 1927)
Anderson, Verity, *Friends and Relations: Three Centuries of Quaker Families* (London, 1980)
Archive Teaching Units – see Dean and Gard
Awdry, Christopher, *Encyclopaedia of British Railway Companies* (Wellingborough, 1990)
Bailey, Brian, *George Hudson: The Rise and Fall of the Railway King* (Stroud, 1995)
Bailey, Michael R., 'Robert Stephenson and Company, 1823–1829', *Transactions of the Newcomen Society*, vol. 50 (1980), pp. 109–37
Barber, B., 'The Concept of the Railway Town and the Growth of Darlington 1801–1911: A Note', *Transport History*, vol. 3 (1970), pp. 283–92
Booth, Henry, *An Account of the Liverpool and Manchester Railway* (Liverpool, 1830)
Briggs, Asa, *Victorian Cities* (Harmondsworth, 1968)
Brooke, David, 'The Promotion of four Yorkshire railways and the share capital market', *Transport History*, vol. 5, no. 3 (1972), pp. 236–67
Clapham, Sir John, *An Economic History of Modern Britain: The Early Railway Age, 1820–1850* (Cambridge, 1926)
Conder, F.R., *Personal Recollections of English Engineers and of the Introduction of the Railway System in the United Kingdom* (London, 1968)
Cottrell, P.L., and Ottley, G., 'The Beginnings of the Stockton and Darlington Railway, 1813–25: A Celebratory Note', *Journal of Transport History*, new series, vol. 3, no. 2 (1975), pp. 86–93
Dale, Rodney, *Early Railways* (London, 1994)
Davies, Hunter, *A Biographical Study of the Father of the Railways* (London, 1975)
Davies, Randall, *The Railway Centenary, A Retrospect* (L&NER, London, 1925)
Dean, S.C., and Gard, R.M. (eds), *Archive Teaching Unit No. 11: The Stockton and Darlington Railway 1825* (University of Newcastle upon Tyne School of Education, 1975)
Dendy-Marshall, Chapman Frederick, *A History of Railway Locomotives down to the End of the Year 1831* (London, 1953)
Dixon, Waynman, *Intimate Story of the Origins of Railways – By 'W.D.'* (Darlington, 1925)
Edwards, K.H.R., *Chronology of the Development of the Iron and Steel Industries* (Wigan, 1955)
Ellis. C.H., *Railway Carriages in the British Isles* (London, 1965)
Flynn, George J., *Darlington in Old Picture Postcards*, vol. 1 (Darlington, 1983)
——, *Darlington in Old Photographs: A Second Selection* (Stroud, 1992)
——, *Darlington in Old Picture Postcards*, vol. 2 (Darlington, 1994)
Gernsheim, Helmut and Alison, *A Concise History of Photography* (London, 1965)
Glass, John, 'The Opening of the Shildon Tunnel', *North-Eastern Railway Magazine*, vol. 3, no. 29 (May 1913), pp. 103–4
Guyonneau de Flambour, F.M., *Practical Treatise on Locomotive Engines on Railways* (London, 1835)
Harrison, J.K., and Harrison, A., 'Saltburn-by-the-Sea: The Early Years of a Stockton and Darlington Railway Company Venture', *Industrial Archaeology Review*, vol. 4, pt 2 (1980), pp. 135–59
Hawkes, R.G.R., *Railways and Economic Growth in England and Wales, 1840–70* (London, 1970)
Hayes, R.H., and Rutter, J.G., *Rosedale Mines and Railway* (Scarborough, 1974)
Head, George, *A Home Tour through the Manufacturing Districts of England in the Summer of 1835* (London, 1836)
Heavisides, M. (ed.), *The History of the First Public Railway, Stockton and Darlington: The Opening and What Followed* (Stockton-on-Tees, 1912)
Holmes, P.J., *The Stockton and Darlington Railway, 1825–1975* (Ayr, 1975)
Hoole, K., *A Regional History of the Railways of Great Britain*, vol. 4, *The North East* (Newton Abbot, 1965)
——, *The Stainmore Railway* (Clapham, 1973)
——, *Anniversary Celebrations of the World's First Steam-Worked Public Railway* (Clapham, 1974)

Hughes, Geoffrey, *London and North Eastern Railway* (London, 1987)
Isichei, Elizabeth, *Victorian Quakers* (Oxford, 1970)
Jeaffreson, J.C., and Pole, William, *The Life of Robert Stephenson* (2 vols, London, 1864)
Jeans, James Stephen, *Jubilee Memorial of the Railway System: A History of the Stockton and Darlington Railway and a Record of its Results* (London and Newcastle upon Tyne, 1875 and 1975)
Joy, David, *A Regional History of the Railways of Great Britain*, vol. 14, *The Lake Counties* (Newton Abbot, 1983)
Kenwood, A.G., 'Transport Capital Formation and Economic Growth on Teesside, 1820–50', *Journal of Transport History*, vol. 11, pt 2 (1981), pp. 53–71
Kidner, R.W., *The Early History of the Locomotive, 1804–1876* (London, 1956)
Kirby, Maurice W., *Men of Business and Politics: The Rise and Fall of the Quaker Pease Dynasty of North-East England, 1700–1943* (London, 1984)
——, *The Origins of Railway Enterprise: The Stockton and Darlington Railway, 1821–1863* (Cambridge, 1993)
Larkin, Edgar J. and John G., *The Railway Workshops of Britain, 1823–1986* (London, 1988)
Lewin, Henry Grote, *Early British Railways: A Short History of their Origins and Development, 1801–1844* (London, 1925)
Lillie, W., *The History of Middlesbrough* (Middlesbrough, 1968)
Lloyd, Christopher, *Memories of Darlington*, vol. 1 (Darlington, n.d.)
——, *Memories of Darlington*, vol. 2 (Darlington, n.d.)
Longridge, Michael, and Birkinshaw, John, *Remarks on the Comparative Merits of Cast Metal and Malleable Iron Railways, and An Account of the Stockton and Darlington Railway* (Newcastle, 1827)
Longstaffe, William Hylton Dyer, *The History and Antiquities of the Parish of Darlington* (Darlington, 1854 and 1973)
Macnay, Charles, *George Stephenson and the Progress of Railway Enterprise* (Newcastle upon Tyne, 1881)
McCord, Norman, *North-East England: The Region's Development 1760–1960* (London, 1979)
McDougall, C.A., *The Stockton & Darlington Railway 1821–1863* (Darlington, 1975)
Mercer, Stanley, 'Trevithick and the Merthyr Tramroad', *Transactions of the Newcomen Society*, vol. 26 (1947–9), pp. 89–103
Mewburn, Francis, *The Larchfield Diary: Extracts from the Diary of the Late Mr Mewburn, First Railway Solicitor* (London and Darlington, 1876)
North, G.A., *Teesside's Economic Heritage* (Cleveland, 1975)
Oeynhausen, Carl von, and Dechen, Heinrich von, 'Report on Railways in England in 1826–27', trans. E.A. Foward, *Transactions of the Newcomen Society*, vol. 29 (1953), pp. 1–12
Ord, J.W., *The History and Antiquities of Cleveland* (Middlesbrough, 1846)
Pangbourne, Joseph Gladding, *The World's Railway: Historical, Descriptive and Illustrative* (New York, 1894)
Pearce, T.R., *The Locomotives of the Stockton and Darlington Railway* (London, 1996)
Pearce, T., and Eden, T., 'The Black Boy Branch', *The Locomotive, Railway Carriage and Wagon Review* (June 1925)
Pease, Mary H., *Henry Pease: A Short Story of His Life* (Darlington, 1897)
Pease, Sir Alfred E., (ed.), *The Diaries of Edward Pease* (London, 1907)
Priestley, Joseph, *Historical Account of the Navigable Rivers, Canals and Railways Throughout Great Britain* (London, 1831)
Proud, John H., *The Chronicle of the Stockton & Darlington Railway to 1863* (Hartlepool, 1998)
Raistrick, Arthur, *Quakers in Science and Industry* (Newton Abbot, 1968)
Raynes, D.H., and Hemingway, J.E., (eds), *The Geology and Mineral Resources of Yorkshire* (York, 1974)
Reed, Brian, *Locomotion. Loco. Profile No. 25* (London, 1972)
Reid, H., *Middlesbrough and its Jubilee* (Middlesbrough, 1881)
Richardson, Thompson, *History of the Darlington and Barnard Castle Railway* (London, 1877)
Robbins, Michael, *The Railway Age* (Harmondsworth, 1965)
——, *George and Robert Stephenson* (London, 1981)
Rolt, L.T.C., *The Cornish Giant: The Story of Richard Trevithick, Father of the Steam Locomotive* (London, 1960)
——, *George and Robert Stephenson: The Railway Revolution* (London, 1960)
Rounthwaite, T.E., 'An Outline History of the Stockton & Darlington Railway', *Railway Observer*, vol. 26 (1956), pp. 14–27
——, *The Railways of Weardale* (Railway Corresponding and Travel Society, 1965)
——, 'The Black Boy Branch', *North-Eastern Express* (February–March 1969)
Russell, Alan, *Yarm and the First Railway* (Yarm, 1990)
Semmens, Peter W.B., *Exploring the Stockton and Darlington Railway* (Newcastle upon Tyne, 1975)
——, *Stockton and Darlington: 150 Years of British Railways* (London, 1975)

Simmons, Jack, (ed.), *Rail 150: The Stockton and Darlington Railway and What Followed* (London, 1975)

——, 'Rail 150: 1975 or 1980?', *Journal of Transport History*, vol. 7, no. 1 (1980), pp. 1–8

——, *The Victorian Railway* (London, 1991)

Smiles, Samuel, *The Story of the Life of George Stephenson* (London, 1859)

——, *Lives of the Engineers*, vol. 3, *George and Robert Stephenson* (London, (1862), Popular Edition 1904)

Speakman, Lydia, and Chapman, Roy, *Stockton on Tees: Birthplace of Railways* (Hawes, 1989)

Stephenson, Robert, and Locke, Joseph, *Observations on the Comparative Merits of Locomotives and Fixed Engines as Applied to Railways, Compiled from the Reports of Mr George Stephenson* (Liverpool, 1830)

Stretton, Clement Edwin, *The Locomotive Engine and its Development* (London, 1892)

Strickland, William, *Reports on Canals, Railways, Roads and Other Subjects, Made to the Pennsylvania Society for the Promotion of Internal Improvement* (Philadelphia, 1826)

Suddes, Alan, *A Grand Tour: John Dobbin (1815–1888)* (Darlington, 1996)

Temple, John, *Darlington and the Turnpike Roads* (Darlington, 1971)

Tomlinson, William Weaver, *The North Eastern Railway: Its Rise and Development* (Newcastle upon Tyne and London, 1915)

Tuffs, Peter, *Eston and Normanby Ironstone Mines* (Middlesbrough, 1996)

Walker, James, and Rastrick, J.U., *Liverpool and Manchester Railway. Report to the Directors on the Comparative Merits of Locomotive and Fixed Engines as a Moving Power* (Liverpool, 1829)

Warren, J.G.H., *A Century of Locomotive Building by Robert Stephenson and Company 1823–1923* (Newcastle upon Tyne, 1923)

West, Theodore, *An Outline of the Growth of the Locomotive Engine, and a few Early Railway Carriages* (London, 1885)

——, *The Evolution of the Locomotive Engine* (Darlington, c. 1900)

——, 'Types of Railway Locomotive Engines on the Stockton and Darlington and the North Eastern Lines, 1825–1848', *Jubilee of the Railway News* (1914)

Whishaw, Francis, *The Railways of Great Britain and Ireland, Practically Described and Illustrated* (London, 1840)

Whittle G., *The Railways of Consett and North West Durham* (Newton Abbot, 1971)

Young, Robert, *Timothy Hackworth and the Locomotive* (London, 1923)

Index

A1 Steam Locomotive Trust, 3
Abercynon Wharf, 36
Action, 206
Adamant, 56
Adamson, Daniel, 54–5
Adamson, Revd. James, 55, 67
Adelaide Colliery, 110, 128, 136
Albert Hill, Darlington, 80, 107, 111, 144–6, 156–7, 159–60
Albion, 56
Allen's Curve, 148
Allen, George, 2
Allens West Station, 46, 109
Allhusen, C., 139
Anniversaries & Celebrations, ix, 51, 96, 152–3, 165–8
Appleby, 137
Auckland Coalfield, 3, 5–7, 16, 56, 59, 114, 124, 134, 140, 143, 148–9, 161
Auckland Park Colliery, 128
Aycliffe (Road, Lane), x, 23, 41, 67, 73, 75–6, 110, 165, 175

Backhouse, Edward, 87
Backhouse, John, 8
Backhouse, Jonathan, 4, 6, 8–9, 12, 14, 17, 111, 114
Bailey, Edward Hodges, 172
Baltimore & Ohio Railway, 62
Bankfoot, 21, 132
Bank Top, Darlington, 155–6, 160–1, 164
Bank Top Station *see* Darlington
Barber, B., 154–5
Barnard Castle, 1, 93–4, 132, 134–6, 138
  East Junction, 135–6
  West Junction, 135–6
  Station, 135, 138
Barras, 134–5
Barrington, Viscount, 8
Barrow in Furness, 126, 133–4, 136
Beaumont, Elizabeth, 87
Bedlington Ironworks, 25, 101
Beechwood, 161
Belah Viaduct, 134–5
Bell Brothers Ironworks, 150
Belle View, 19
Bell, Isaac Wilson, 150
Bell, John, 150
Bell Rock Lighthouse, 5
Bell, Sir Isaac Lowthian, 150
Bell, Thomas, 150
Bell, William, 157
Belmont, 142
  Farm, 151
  Junction, 151
  Mine
Berwick & Kelso Railway, 7
Birkbeck, Henry, 115, 119
Birkinshaw, John, 25–6
Bishop Auckland, 20, 23, 56, 91, 94–5, 123–26, 128–32, 134–6, 147
Bishop Auckland & Weardale Railway, 33, 126, 128, 130–1, 133–4, 136, 150
Bishopley, 11
Black Boy Branch, 33, 78, 102, 108–11, 127–8
Black Boy Colliery, 33, 110–11, 115, 128

Black Callerton Colliery, 204
Black Diamond, 74, 98, 107, 166
Blackett, Christopher, 36–8
Blackhill, 130
Blansard, Henry, 149
Blenkinsop, John, 37, 77
Blucher, 13, 37
Bolckow, Henry William Ferdinand, 139
Bolckow & Vaughan, 138–9, 150
Bolton & Leigh Railway, 102
Bonomi, Ignatius, 47, 54
Bonomi, Joseph, 47
Boosebeck, 139
Booth, Henry, 57
Borrowses, 87
Botcherby, Robert, 8
Botcherby, Thomas, 4
Bouch, Thomas, 130, 134–5, 137
Bouch, William, 80, 103, 195, 134–5, 153, 169
Boulton & Watt, 61, 103
Bowes, 135–6
Bowesfield, 114, 116, 148
Boyne, Lord, 131
Bradbury, 3
Brancepeth Colliery, 195
Brandling Junction Railway, 32
Brandon Colliery, 195
Brandreth, Thomas Shaw, 72–3
Brindley, James, 1
Brinkburn, 161
Brook, Henry, 127
Brougham, 103, 135
Brown, Captain Samuel, 116
Brown, J.R., 54–5
Brunel, Isambard Kingdom, 116
Brusselton, 75
  Colliery, 21, 113
  Inclined Planes, 16, 21, 23, 26, 28–33, 35, 43–4, 63, 78, 98, 113, 130–2
  Quarry, 26
  Ridge, 20, 23, 30
  Summit, 23, 30–1, 33, 43–4, 96, 110
  Tower, 43

Cairns, Jeremiah, 3, 4, 6
Canals, 1–5, 7, 10, 26, 58–9
Canterbury & Whitstable Railway, 19, 73
Cargo Fleet, 1, 151
Carr House, 33, 129–30
Carriages/Coaches, 91–3, 99, 170–1
Cartagena, 36
Castell, John, 128
'Catch-me-who-Can', 36
Centenary, ix, 167, 169, 173, 175
Charity Junction, Darlington, 133
Chat Moss, 23
Chauldron Wagons, x, 14, 27, 33, 37, 44–5, 50, 71–4, 76, 83, 93, 99, 110–11, 161
Chaytor, William, 6, 8
Chesterfield., 176–7
Chilton, William, 115
Chittaprat, 101
City of Durham Branch, 149
Clarence, Duke of, 149
Clarence Railway, 23, 111, 148–50
Clark, Thomas, 166

Cleveland, 122, 126, 133, 138–40, 150–1
  Hills, 120, 139, 142
  Ironworks, 140
Cleveland, 202
Cleveland, Dukes of, 6, 132, 134, 161, 171
Cleveland Railway, 150–2
Cleveland Railway Extension, 151–2
Clifton, 137
  North Junction, 137
Close, Thomas, 66
Coal Leaders, 74, 82–3
Coatham, 140
Cockermouth, 137
Cockermouth, Keswick & Penrith Railway, 137–8
Cockermouth & Workington Railway, 137
Cockerton, 1–2
Cockfield Fell, 113
Cod Hill Mine, 142
Colombian Mining Association, 40
Compulsory Purchase, 24.
Consett, 123, 125–6, 128–30, 133, 138
Consett Iron Company, 201
Contractor, 105
Copley Bent Colliery, 112–13
Cottage Row, Stockton, 23, 89
Coulthard, Thomas, 83
Coundon, 110
  Gate Colliery, 110–11
  Grange Colliery, 110
  Jawblades Colliery, 199
  Turnpike Gate Branch, 108, 110
Coxon, John, 91
Crawley, 33, 129
Creed, William, 197
Cree, John, 75
Crewe, 15, 167
Croft, 1, 4, 15, 94, 111, 124, 143
  Bridge, 111, 143
  Depot Branch, 16, 17, 80, 108–9, 111–12, 114, 124, 144, 147
  Junction, 144, 147
Crook, 33, 93–4, 128–31, 150, 167, 169
Crumlin Viaduct, 135
Cubitt, William, 119
Cuneo, Terence, 48

Dale, Sir David, 105
Darlington
  Bank Top Station, 123, 146–7, 155–60, 162–4, 167–8
  Coal Depot Branch, 42, 46, 55, 91, 105, 108–9, 124, 161
  Coat of Arms, 127, 154, 158, 162–3
  Committee, 153
  Earl of, 6–7
  Forge, 125
  Junction, 73
  Library, 155–6
  Mechanics Institute, 155–6
  North Road Station, 46, 55, 95–7, 105, 109, 126, 132, 144–5, 147, 153, 155, 160, 167–8
  Railway Institute, 156
  Town Hall, 5, 34, 155, 175
Darlington, 159, 163

INDEX

Darlington & Barnard Castle Railway, 132–3, 138
Darlington Locomotive Works, 104–5 (*see also* North Road Shops)
Darlington Railway Centre & Museum, 52, 55, 96, 165, 168–70
Darlington Railway Preservation Society, 97
Dart, 103
Davies, David, 4, 7
Deanery Branch Railway, 200
Deanery Colliery, 110, 128
Dechen, Von. H., 32
Deepdale Viaduct, 134–5
Deerness Branch, 33, 131
Deerness Valley Railway, 131
Defence, 67
Defiance, 67
De Montgolfier, Mgr., 57
Dene Beck, 110
Dennies, Thomas, 127
Derwent, 106, 167–9
Derwent Iron Company, 125, 128–9, 138
Derwent
 Railway, 129
 Valley, 133
Dewley Burn Colliery, 10
Dickinson, George, 121
Diligence, 74, 98, 163

Dinsdale, 125
Dinsdale Spa, 94, 147
Director, 166
Director locomotive class, 101, 159, 169
Dixon, George, 1, 19
Dixon, James, 116
Dixon, John, 12, 19–20, 46, 84, 104, 112, 176
Dobbin, John, 51–4, 62, 155, 168
Dobson, Joseph, 169
Dodds, Ralph, 38
Dorman Long, 150
Double-acting drums, 29–30, 100
Downing, Nicholas, 103, 111
Dresser, William, 146
Durham City, 12, 147, 149
Durham & Cleveland Union Railway, 150–1
Durham Junction Railway, 147

Eaglescliffe (Egglescliffe), 23, 47, 109, 148
Eamont Bridge Junction, 137
Eastgate, 132
East Mount, 86
East Mount, 161
Eclipse, 198
Eden Valley, 138
Eden Valley Junction, 137
Eden Valley Railway, 137
Edge Rails, 13, 18, 37
Edward Pease, 163
Eldon, Earl of, 6–7
Eldon Pits, 110, 128
Ellens, The, 86
Elm Field, 161
English Electric, 160
Enterprise 105, 169
Eston
 Hills, 138–41, 150–1
 Junction, 140
 Mines, 140
Etherley, 131
 Colliery, 4
 Inclined Planes, 16, 21, 28–31, 33, 35, 43, 130–1
 Quarry, 26
 Ridge, 20–1, 30, 112
Etty, William, 156
Euston Square, London, 36, 172
Euston Station, 172–3

Evenwood Lane, 108, 112–13
Exmouth, 94
Experiment coach, 44, 55, 59–63, 65–6, 71, 82, 88, 90–1, 169, 174
Experiment locomotive, 77
Express, 66

Fares, 63, 88, 90, 93
'Father of Railways', 1, 10, 14, 51, 94, 99, 141, 172, 176
Faverdale, Darlington, 167, 174
Feethams, 87, 161
Ferryhill, 150
Fieldon Bridge Junction, 132, 136
Fighting Cocks, 23, 67, 73, 88–9, 94
Fish-bellied rail, 25, 111
Flatow, Victor Louis, 198
Fleece Inn, Stockton, 66
Flounders, Benjamin, 1, 3, 6, 20, 114
Flying Scotsman, ix, 167, 175
Forcett Railway, 171
Forth Banks, Newcastle, 35, 159
Forth Street Works, Newcastle, 30, 35, 40–1, 77, 98, 100–1, 159
Fossick, Sara, 87
Foster, Jonathan, 37
Fowler, Marshall, 23
Fox-Davis, Arthur, 162
Fox, Mary, 141
Freeholders' Estate, Darlington, 161
Friends, Society of, 155, 176
Frosterley, 130
Frosterley & Stanhope Railway, 131, 137
Furness Railway, 134, 153

Gainford, 132–3
Cateshead-on-Tyne, 144, 147
Gauge, 27
Gaunless River, 20–1, 43, 112–13, 131–2, 136
George Stephenson & Son, 19
Gibson, Francis, 115
Gilkes, Edgar, 119
Gilkes, Oswald, 103, 111
Gilkes, Wilson & Co., 103, 119
Gillespie, John, 75
Gladstone, W.E., 120–1
Globe, The, 78–9, 116
Goosepool, 23, 47, 49, 51, 67, 73, 109
Gowland, George, 101
Gowland, James, 89, 102
Gowland, William, 43, 102
Graham, George, 75–6, 101
Graham, John (Railwayman), 70, 79
Grainger, Thomas, 148
Grand Allies, 11, 37, 58
Grand Junction Railway, 19
Grangetown, 140
Great Ayton, 115, 142
Great North of England Railway,
Great North Road, 23, 46, 96–7, 154, 156
Great Stainton, 14
Great Western Railway, 80, 155
Greener, Thomas, 33
Greenfield, 32
Greta, River, 135
Grimshaw, John, 6
Grove, The, 83
Guisborough, 139–41, 150–1
 Goods Station, 142
 Junction, 141
 Priory, 141
Gurney, Emma, 9, 86
Gurney, Joseph, 6, 8–9, 86, 115
Gurney, Joseph John, 8
Gurney Pit, 110, 128
Gurney, Samuel, 9

Hackworth, 163
Hackworth, John Wesley, 63, 103
Hackworth, Thomas, 103
Hackworth, Timothy (1786–1850)
 Appointed S&DR Resident Engineer, 30; improvements to stationary steam engines and inclines, 30; invents 'double-acting' drums, 29, 30–1; advocacy of double-track inclines, 31–2; invents 'dog' discharge hook and 'cow' drag frame, 32–3; correspondence with Edward Pease, 30–3; birth and early employment, 41; at Forth Street Works, 41; and *Locomotion*, 41; at Inaugural Run, 44; opinion of passenger-coach drivers, 69; invents spring safety-vale, 75; designer of *Royal George*, 77–8, 101; designer of *Sans Pareil*, 100, 102; designer of 'Majestic' and 'Director' classes, 101; appointed Locomotive Superintendent, 99; and Liverpool Deputation, 99; invents plug wheel, 100–1; founds Soho Engine Works, 100, 102; benefactor, 100; statue, 100; as private contractor to S&DR, 103; death, 103; son John Wesley succeeds, 103; home, 103–4; designs Middlesbrough coaling staithes, 115; locomotive named after, 163
Haggerleases (Lane) Branch, 42, 108–9, 136
Haggerleases Station, 113
Hardwicke, Phillip Charles, 172
Harris, John, 161
Harrison, T.E., 129
Hartburn Junction, 148
Hartlepool, 111, 119
Haverton Hill, 114, 149
Hawkshaw, John, 151
Hawthorn, Robert, 159
Hawthorn, William, 159
Heaviside, M., 23
Hedley, William, 37, 99–100
Heighington, 20, 23, 41, 110
Hetton Colliery, 12, 27, 70–1
High Pit Colliery, 11, 37
High Shildon, 33, 110
High Street House, 10, 36
Hill, Anthony, 36
Hill, Ralph, 69
Hincaster Junction, 134
Hodgson, George, 69
Hollyhurst, 161
Holmes & Pushman, 26
Homfrey, Samuel, 36
Hope, 35, 74–5, 98
Hopetown, 105, 132, 155–6, 160
 Carriage Works, 171
 Junction, 105
 Lane, 105, 109
Hopkins, John Castell, 192, 201
Horner & Wilkinson, 170
Howne's Gill, 33, 129–30
Howden incline, 33
Howden Station, 167
Howsham, 145
Howson, Martha, 67
Hudson, George, 144–5, 147, 151, 156, 173
Hummersknott, 161
Hunswick Colliery, 195
Hurworth Place, 111
Hutton Gate, 141–2
Hutton Junction, 151
Hutton Lowcross, 141–2
Hutton Village, 142

Ianson, Joshua, 111

INDEX

Inclined Planes, 16, 21, 27–33, 43, 58, 64, 110, 129–30
Industrial Revolution, xi
Ingleby, 139
Institute of Mechanical Engineers, 176
Ixion, 203

Jackson, Ralph Ward, 150–1
Jackson, Thomas, 20
James, William, 18, 39
Jeans, J.S., x, 8
Jessop, William, 25
John Dixon, 163
Johnson, William, 90
John Warne & Sons, 158
Josiah Wedgwood & Sons, 173
Jowett, Joseph, 86
Jowett, Sophia, 86
Joyce, Harry, 102
Jubilee, ix, 51, 166–7

Kendal & Windermere Railway, 201
Keswick, 137
Killingworth Colliery, 11–13, 27, 36–7, 42, 85
Killingworth engine, 12–14, 35, 37, 39–40, 85
King
    George III, 7
    George IV, 7, 10, 112
    William IV, 149
    Edward VII, 127
    George VI, 174
King's Head Inn, Darlington, 2, 14, 67, 87, 146
Kips, 32
Kirby, Maurice, viii, 2, 9, 24, 38, 103
Kirkby Stephen, 134–5, 137–8
Kitching, Alfred, 31, 121, 132
Kitching, John, 6, 8
Kitching, William, 6, 43, 98

Lancaster, 94
Lancaster & Carlisle Railway, 134, 137
Lanchester, J., 45
Lands, The, 113
Larchfield, 15
Larchfield, 161
Lardner, Dr. Dionysius, 38
Lawson, Joseph, 171–2
Leather, George, 3
Leeds Northern Railway, 61, 148, 151, 157–8
Leeds & Thirsk Railway, 202
Liddell, Sir Thomas, 37
Linen industry, 154
Lister, William, 98, 102, 106, 154–5, 166
Liverpool, 167, 172
Liverpool, 197
Liverpool & Birmingham Railway, 19, 49
Liverpool & Manchester Railway,
Liverton, 139
Livery, 92
Locke, Joseph, 33
Locomotion No. 1, 35–6, 40–1, 43–5, 47–8, 51, 53, 55, 74–6, 88, 91, 98–9, 101, 104, 121–2, 162–3, 165–9, 174–5
Loftus, 94, 139
Londesbrough Hall, 145
London & Birmingham Railway, 199
London, Midland & Scottish Railway, 172
London & North Eastern Railway, 1x, 105, 145, 152, 156, 159, 163, 167–8, 173–5
London & North Western Railway, 134, 137, 152, 167, 172
Longridge, Michael, 25–6, 35, 41, 101
Longstaffe, Hylton Dyer, 111, 117
Losh, William, 26, 38
Losh, Wilson & Bell, 26, 35, 38, 139, 150
Lough, John Graham, 173
Lowther, 103, 135
Lowther, Sir H.J., 138

Ludley & Buckden, 71

Macadam, 67
Mach, David, 173
Machine Pit, 110, 128
Majestic, 165–6
'Majestic' locomotive class, 101
Mallard, 173
Malleable rails *see* Wrought Iron
Maltby, Bishop, 15
Mandale Cut, 1
Maria, 115
Marley, John, 138
Marochetti, Baron Carlo, 173
Marshall, James, 87
Marske Station, 140
Martin, Simon, 115
Marton, 139
Masons Arms, 23, 44, 70, 98, 103, 110, 127
Maybey, Charles H., 173
McNay, Alexander, 169
McNay, Thomas, 121, 132, 148
Medal
    Centenary, ix, 174–5
    Middlesbrough Extension, 117
Meeting Slacks, 23
Merrifield, L.S., 173
Metcalfe, Robert, 41, 46, 166–7
Mewburn, Francis, x, 6–8, 10–11, 15–16, 24, 46–49, 55, 62, 114, 121, 126, 128
Meynell, 163
Meynell, Thomas, 3–4, 6–8, 14, 18, 20, 44, 46–9, 55, 62, 114, 121, 126, 128
Middlesbrough & Guisborough Railway, 140–2, 150–1
Middlesbrough, 56, 79, 83, 89, 91, 94–5, 103, 111, 114–20, 125–6, 138–9, 147–51
    Dock, 118–20
    Dock Junction, 120–1
    Station, 119, 121, 141
Middlesbrough Extension Railway, 64, 70, 79, 113, 114–19, 123, 139, 148–9
Middleton Colliery, 32
Middleton Railway, 32, 37
Middleton, John, 156
Middleton St George, 67, 94, 148
Midland Railway, 137
Miles, Richard, 3–4
Milestones, 81
Miller, William, 199
Millom, 126, 133–4, 136
Mires, Robert, 197
Morton Grange, 150
Motto, 14, 53, 154, 168, 176
Mount Pleasant, 33, 148
Murchamp, Emmerson, 127
My Lord, 37
Myers Flatt, 23, 69

Nanny Mayor's Incline, 33, 129–30
National Portrait Gallery, 173
National Railway Museum
Navvies, 58, 135
Neath Abbey Ironworks, 26
New Brancepeth Colliery, 195
New Shildon, 20, 23–4, 42, 44–6, 48, 50, 61, 75, 77, 91, 94, 98–100, 104, 110, 113, 126, 128, 132, 149
Newburn, 11, 37, 11, 37
Newcastle & Carlisle Railway, 133
Newcastle & Darlington Junction Railway, 144, 147, 157
Newcastle & Derwent Railway, 202
Newcastle Literary & Philosophical Institute, 173
Newcastle upon Tyne, 10–12, 21, 26, 31, 41, 43, 62, 79, 102–3, 107, 139, 143–4, 147, 160, 167, 169, 173

Newlandside, 131
Newman, Thomas, 10
Newport, 1
Newton Cap Colliery, 195
Norlees House, 21, 108, 112
Normanby, 150–1
North Eastern Railway, 86–7, 103–5, 111, 129–30, 133, 137, 145, 148, 150–3, 156–8, 167–70
North Lodge, 87
North Midland Railway, 61
North Road Shops, 104–5, 125, 156, 160, 163, 166–9
North Road Station *see* Darlington
North Roddymoor Colliery, 131
North Stockton Station, 148
Northern Lighthouse Board, 5
North Yorkshire & Cleveland Railway, 148
Nunthorpe, 141, 151

Oakenshaw Colliery, 195
Oaktree Junction, 23, 123, 159
Oeynhausen, Von. Carl, 32
Office, S&DR, 20, 87
Office, GNER, 146
Ogle, William, 69
Old Black Boy Colliery, 110
Old Etherley Colliery, 21, 56
Old Town Junction, Middlesbrough, 120
Ormesby, 94, 141
Otley, Richard, 114, 120
Overend, Gurney & Co., 9
Overton, George, 3–7, 12–13, 15–18, 20–1, 23–4, 28, 30, 64, 109, 112, 167
Owners of the Midlesbrough Estate, 115, 119–21

Parcels, 91
Parkgate Junction, Darlington, 123, 144, 146–7
Park Head, 33
Parliamentary Trains, 95
Patter, William, 99
Peachey, William, 121–2
Peacock, Daniel Mitford, 8, 14
Peacock, Thomas, 17
Pearce, T.R., viii, 168
Pearson, Sarah, 91
Pease, Edward (1767–1858)
    'Father of Railways', x, 14, 51, 99, 141, 176; tributes, 4; interests, 4; investment in S&DR, 6, 8–9; and 'Darlington Party', 3, 6; character, 7; and George Stephenson, 10–12; visits Killingworth, 13–14, 35; and Francis Mewburn, 15; recommends George Stephenson as Engineer, 18; instructions to George Stephenson, 18; and George Overton, 24; and Timothy Hackworth, 30–1; and Robert Stephenson & Co., 35, 41; On Inaugural Run trial train, 43; death of son Isaac, 49; indebtedness of S&DR, 49; birthplace, 83; education and marriage, 83; and 'The Ellens', 86; son, Joseph, and brother, Joseph, 86–7; and Liverpool Deputation, 99; and Middlesbrough Extension, 114; as entrepeneur, 139; and the Middlesbrough & Guisborough Railway, 141–2; descendants, 141–2, 204; locomotive named after, 163; proposed statue, 172; portrait on Centenary medal, 175; attends George Stephenson's funeral, 176–7; death and grave, 176
Pease, Edward (1711–1785), 83
Pease, Edward (1801–1839), 43, 115
Pease, Elizabeth, 85

INDEX

Pease, Elizabeth Beaumont, 87
Pease, Emma, 9, 86
Pease, Henrietta, 155
Pease, Henry, 31, 43, 51, 86, 121, 132–3, 135, 142, 152–3, 155
Pease, Isaac, 49
Pease, John, 86, 121, 132, 142, 151–2
Pease, John Beaumont, 87
Pease, Joseph (1737–1808), 83
Pease, Joseph (1772–1846), 87
Pease, Joseph (1799–1872) *see below*
Pease, Mary, 49, 85
Pease & Partners, 140, 167, 169
Pease, Rachel, 83
Pease, Sophia, 86
Pease, Sarah, 87
Pease, Sir Alfred Edward, 11, 13, 142
Pease, Sir Joseph Whitwell, 121, 131–2, 141–2, 155
Pease, Thomas Benson, 8
Pease, Joseph (1799–1872)
  Draws up S&DR Prospectus, 6; investment in S&DR, 17; accident at Brusselton, 31; and George Stephenson, 41; on Inaugural Run trial train, 43; and economics of coal, 56; birth and marriage; succeeds Edward Pease as promoter of the S&DR; evidence to Commons committee, 90; and Soho Works, 103; owner of Adelaide Colliery, 110; and development of Middlesbrough, 114–15; and The Owners of the Middlesbrough Estate, 115; and Shildon Tunnel Co., 126; and Weardale Extension Railway, 128; and Deerness Valley Railway, 131; and Duke of Cleveland, 133; and the South Durham & Lancashire Union Railway, 135; and the Midlesbrough & Guisborough Railway, 141–2; and the Great North of England Railway, 143; opposition to the Newcastle & Darlington Junction Railway, 151; opposition to the Cleveland Railway, 151; statue, 171–2; gift of Darlington Town Clock, 176; grave, 204
Pease's Mill, 34
Pease's West Colliery, 167, 169
Pennines, 134–6, 138
Penrith, 123, 136–7
Pennsylvania Society for the Promotion of Internal Development, 57
Pen-y-Daren, 36
  Ironworks, 36
Perseverance, 70
Phillimore, John, 54
Phoenix Foundry, Shildon, 111
Phoenix Row, 21
Phoenix Pit, 21
Pickersgill, Richard, 8, 41, 66, 71, 88
Piercebridge, 1, 4, 132
Pierremont, 132, 155
Pierremont, 161
Pilot, 144
Pinchinthorpe, 141
  House, 142
  Station, 142
Pipewellgate Foundry, 86
Pitts, Joseph, 173
Planet, 197
Plug wheel, 100–2, 166, 175–6
Polam, 161
Polam Junction, 159
Pontop & South Shields Railway, 129
Port Clarence, 149–50
Port Darlington, 116, 119–20, 149
Portrack, 114

Prattman, Revd. William Luke, 112–13
Preston Junction, 148
Preston Park, 23, 148, 168
Prior, Alan, 168–9
Prpoerty plate, 22, 86, 109, 111, 128
Proprietors, 10
Prospectus, 6, 17, 56, 114
Proud, John, 91
Prudhoe, Lord, 127
Public Railway, xi, 58, 82
Puffing Billy, 37–8

Quaker(s), ix–x, 1, 3, 8–9, 137, 140, 143, 154–5, 161, 171–2, 176
Queen, 105
Quen Elizabeth II, 174–5
Queen Victoria, 145–6, 157, 167, 169

R & W Hawthorne-Leslie & Co., 31, 98, 102, 159, 169
Raby Castle, 105
Railway Mania, 145
Railway time, 80, 94–5
'Railway Times', 76, 144
Rainhill, 113
Rainhill Trials, 79, 100, 102, 162
Rainton Meadows, 147
Raisbeck, Leonard, 1, 3–4, 7–8, 10, 12, 16, 20, 59, 115, 148
Rankley, Alfred, 85
Rastrick, John Urpeth, 31, 36, 114
Redcar, 94, 121, 138–9, 150–1
  Junction, 122
  Station, 122, 140
Redcar & Saltburn Railway, 122, 140
Redhills
  Curve, 137
  Junction, 137
Reliance, 67
Rennie, John, 2–5
Richardson, Mary, 83
Richardson, Thomas, 6, 9, 13, 35, 47, 115
Richmond, 167
Rise Carr, 23
Robert Stephenson & Co., 19, 30, 35, 41, 62, 72, 74–5, 77, 79, 98, 100–1, 103, 107, 134, 159, 168
Robert Stephenson & Hawthorne Ltd., 159–60
Robinson, Robson, 94
Rocket, 100, 102, 166, 175
Roddymoor Colliery, 201
Rolt. L.T.C., viii, 42, 67, 69, 112
Rookhope, 33
Roseberry, 139
Roseberry Topping, 141–2
Rose Cottages, Shildon, 111
Rowlandson, Thomas, 201
Rowley, 129
Rowley Station, 201
Royal George, 77, 79, 100–2, 165–6
Royal George Locomotive Class, 196

Saltburn, 121–3, 138, 167
Saltburn Improvement Company, 121, 138
Samphire Batts, 149
Satow, Mike, 168
Scott, Richard, 67, 71
Scott, Sir Walter, 5, 138
Scrayingham, 145
S&DR Crossing, 107, 125, 145–8
Seal, 14, 34, 175
Seaton, 94
Seguin, Marc, 57, 101
Seguin, Paul, 101
Sesquicentenary, ix, 168
Settle & Carlisle Railway, 137

Share certificates, 34
Shaftesbury, Lord, 16
Sheffield Wednesday, 163
Shildon, ix, 3, 22–3, 42, 63, 69–70, 73, 75–6, 79, 94, 100, 103–5, 132, 165, 167, 171
  Bank Colliery, 110
  Junction, 70, 110, 123, 127, 131
  Locomotive Company, 103–4
  Lodge Colliery, 76
  Parish Church, 100
  Ridge, 23, 110, 126, 132
  Tunnel, 126–7, 131–2
  Works, 98, 102–4, 119, 154, 156
Simpasture, 23, 45, 73, 149
Simpson, John, 91
Simpson, Mary, 96
Skelton, 139
Skelton Beck, 121–2, 150
Skerne Bridge, 47, 54–5, 86, 155, 175
Skerne, River, 3, 23, 47, 53–4, 83, 86, 111
Skew Arch Bridge, 113
Skinningrove, 139–40, 150–1
Smardale Viaduct, 135
Smiles, Samuel, 11, 27, 39, 60–1, 66, 85–6, 148, 166, 169, 176
Smith, Henry Pascoe, 192
Smith, Ralph, 198
South Church, 126, 128, 147
South Durham Colliery, 110
South Durham Ironworks, 125
South Durham & Lancashire Union Railway, 133–8
South Durham Railway, 128
South-Eastern Railway, 61
Southend, 87, 132, 171
Southend, 161
South Shields, 129
South Stockton Station, 148
South-West Durham Coalfield *see* Auckland Coalfield
Sparkes, J.M., 55
Spencer, George, 51
Spennymoor, 199
Springfield, Darlington, 159
Spring Gardens Junction, 134, 136
Stage wagon, 53
Stainmore, 133–5, 137–8
Standalone, 23
Stanhope, 33, 129, 131–2, 168
Stanhope & Tyne Railway, 33, 128–9
Stanley, 33
Stanley Colliery, 131
St. Cuthbert's Church, Darlington, 15, 34, 51, 55, 87, 155
Steele, John, 36
Stephenson, 163
Stephenson, Elizabeth, 177, 204
Stephenson, Fanny, 10–11
Stephenson, George (1781–1848)
  'Father of Railways', x, 10, 94; birth and marriage, 10–11; estimates cost of S&DR, 9; early life and employment, 10, 36; and Killingworth Colliery, 11; meets Edward Pease, 11–12; his 'Killingworth engines', 12, 14, 37–9; agrees to survey S&DR, 15; as prophet, 14; advocates locomotive v horse traction, 17; appointed Engineer to S&DR, 18, 0; sets up office, 19; survey report, 20; finalises route of S&DR, 20; designs Gaunless Bridge, 21; pioneers 'standard gauge', 27; engineers and powers inclined planes, 7; cuts first sod, 23; locomotive designer and builder, 5–39; meets Richard Trevithick, 36; & *Locomotion*, 41; brothers, 42; completion and Opening of S&DR, 42; designs Skerne

INDEX

Bridge, 47; tributes, 49; biography, 60; and dandy cart, 73; and *Experiment*, 77; teaches daughters of Edward Pease, 85; recommends Timothy Hackworth, 100; and *Rocket*, 102; deputises surveying to son Robert 112; consulting engineer to Middlesbrough Extension, 115; locomotive named after, 163; Centenary, 167, 169; statues, 172–3; busts, 173; medal, 175; portrait, ix, 175; retires to Tapton House, Chesterfield, 176; death and interment, 176; memorials in Chesterfield parish church, 176; obituary, 177

George Stephenson and Son, 19, 194
Stephenson, James, 42–3, 165
Stephenson, John, 42
Stephenson Locomotive Society, 55
Stephenson, Robert *see below*
Stephenson, Robert (Snr.), 10
Stevens, John, 48
Stevenson, Robert, 3, 5, 7, 25
Stevenson, Robert Louis, 5
St Helen's Auckland, 43, 83, 91, 94–5, 113
St Helen's Colliery, 132, 136, 195
Stephenson, Robert
    Birth, 11; tribute, 19; and Robert Stephenson & Co., 30; 'Observations on Fixed Engines', 33; meeting with Richard Trevithick, 36; absence abroad, 40, 112; relationships with father George Stephenson, 41; correspondence with Edward Pease, 41; biography, 61; correspondence with Michael Longridge, 75; criticises impediments to locomotive haulage, 76; and *Rocket*, 102; and Haggerleases Branch, 112; gives evidence before Parliamentary Committee, 112; designs Tees Bridge, 117; consulting engineer to Stanhope & Tyne Railway, 129; promotes Pontop & South Shields Railway, 129; and the Great North of England Railway, 143; Engineer of Newcastle & Darlington Junction Railway, 144; locomotive named after, 163; statue and busts, 173; builds first Italian railway, 174
Stillington, 149
St John's Well, Stockton, 20, 23–4
Stobart, Henry, 126
Stockton, 163
Stockton & Darlington (North Riding) Railway, 122
Stockton & Hartlepool Railway, 119, 150
Stockton on Tees
    Branch, 108
    Port, 1, 3, 143
    Quay, 48, 56, 71, 77, 89, 114, 123
    Station, 148
    St John's Crossing, 88, 116, 124
    Town Hall, 1, 3, 20, 49, 95
Stooperdale Curve, 133
Stooperdale Junction, 133
Storey, Francis, 8
Storey, Thomas, 19–20, 71, 77, 99, 113–14, 144, 166
Strathmore, Earl of, 112
Stretton, Clement Edwin, 168
Strickland, William, 32, 39, 57
Suddes, Alan, 52–3, 160
Summerside, Thomas, 42
Sunderland, 145
Sunniside, 115
Sunniside Incline, 33, 130
Surtees Railway, 70–1, 102, 127
Suspension bridge, 116–17
Swin bridge, 113

Swindon, 154

Tapton House, Chesterfield, 176
Tate, William, 8
Tay Bridge, 134–5
Taylor, John, 41
Tebay 123, 134–7
Tees
    Conservancy Commissioners, 120–1
    Engine Works, 119
    Navigation Company, 1, 114, 116
    River, 1, 7–8, 26, 56, 64, 94, 111–12, 114, 116, 133, 135, 138, 148–51, 161
    Valley, 138
Teesside, 39, 124–5, 133–6, 138, 140
Tees Viaduct, 135
Tees & Weardale Railway, 114, 148
Telegraph, 80
Telford, 67
Tennant, Christopher, 3, 114, 143, 149–50
Tennison, Charles, 75
Thickley, 3, 149
Thomas, W., 85
Thompson, Richard, 91
Thornaby, 117, 148
Tickets, 44, 46, 50, 66, 87–90, 94
Timetables, 60, 66, 83, 88, 90–1, 93–6, 110, 113, 135
Timothy Hackworth Victorian & Railway Museum, 100, 103–4, 171, 194
Tinley, Alan, 195
Tolls, x1, 63–5, 67, 82, 111
Tomlinson, William Weaver, 24, 33, 62, 138, 143, 151
Tonnage, 64, 71, 108
Tornado, 105
'Tory' locomotive class, 169
Tow Law, 125, 130, 142, 151
Trader, 169
Train, 173
Trevithick, Richard, 35–8, 77, 173
Troutbeck, 137
Tully, Mr., 66, 88
Tunnel Branch, 131–2, 136
Tunnel North Junction, Shildon, 128
Turnbull, Edward, 75
Turnpikes, 4, 7, 23, 43, 46–8, 51, 58–60, 67, 96
Tyne, River, 10, 36, 39, 40–1, 100, 129, 144, 147

Ulverstone, 133, 136
Union, The (carriage), 92
Union, The (coach), 67, 68, 76, 88, 90–2
Unthank, John, 9
Upleatham, 138–9
Upsall Grange, 151
Urlay Nook, 23, 67, 73, 88

Vane, Henry, 132, 171
Vaughan, John, 138–9
Victoria Road, Darlington, 157

W & A Kitching Ltd., 102–3, 105–6, 144, 154
Walbottle Colliery, 37, 41, 99–100
Walker, James, 31
Walker-on-Tyne, 38, 139, 150
Wandless, Luke, 127
Waskerley Park, 128–9
    Junction, 129
    Reservoir, 169
Waterhouse, Alfred, 175
Waterhouses Colliery, 195
Waterhouses, 131
Watson, D.H., 61
Watt, James, 39, 61
Watts, Captain, 144
Weardale, 125, 129, 131–2, 168

Weardale Extension Railway, 133, 128–30
Weardale Iron Company, 33
Wear & Derwent Junction Railway, 129
Wear & Derwent Railway, 128–30
Wear, River, 8, 20, 23, 36, 40–1, 131
Wear Valley Complex, 126–31, 138
Wear Valley Railway, 130–1
Wear Valley Extension Railway, 201
Wear Valley (Witton) Junction, 130–1
Weatherhill, 33, 129, 131
Weatherills Junction, 137
Wellington, 37–8
Wellington, Duke of, 145
West Auckland, 6, 16, 21, 43, 59, 131–2, Junction, 132
West Durham Railway, 150
West Hartlepool, 119
West Hartlepool Harbour & Railway Co., 150
West Lodge, 161
West Moor Colliery, 11, 13, 36–7, 42
West Stanley, 195
West, Theodore, 60, 168
Wharton, William Lloyd, 111
Wheatstone, Charles, 80
Whessoe, 23, 53–4, 105
Whiley Hill Farm, 23
Whinfield, John, 36
Whitby, 94
Whitehall Junction, 129
Whitehaven, 137
Whitelaw, William, 174
Whiteleahead, 128
Whitfield Gardiner, 129
Whitley Springs, 23, 47, 148
Whitwell, Rachel, 83
Whitworth, Robert, 1–2
Whorlton, 139
Wilberforce, 169
William Lister & Co., 98, 102, 106, 154–5, 166
Willington, 10–11, 27
Wilson, Isaac, 119, 121
Wilson, James, 113
Wilson, Robert, 101
Winding engines, 21, 28–33, 43–4, 103, 110, 129, 131, 175
Winston, 1–2
Witton Park Colliery, 7–8, 20–1, 26, 43, 112, 123–4, 131, 165
Wolferton, 154
Woodhouse Close Colliery, 132
Woodlands, 161
Wood, Nicholas, 11–12
Wooley Colliery, 131
Wordsworth, William, 137, 201
Workington, 126, 136–8
Wrought iron (malleable), 18–9, 22, 25–6, 115
Wylam, 38, 41, 99–100, 175
    Colliery, 36–8, 41, 99–100
    Wagonway, 7
Wylam Dilly, 37–8
Wyon, Edward William, 173

Yarm, 1, 3–4, 6–8, 20, 46–7, 49–51, 67, 114, 121, 124–5, 147–8
    Coal Depot Branch, 50–1, 90, 108–9, 111–12, 114, 124, 148
    Junction, 109–10
    Station, 109, 148
York, 86, 143–5, 152, 156, 158
    Junction, 80
York Herald, 143
Young, Robert, 31
York & North Midland Railway, 145, 152
York, Newcastle & Berwick Railway, 148

Zetland Hotel, 122
Zetland, Lord, 122, 138